SHAPE IT!

**Magnificent
Projects
For Molding
Materials**

U.S. edition published in 1999 by Lerner Publications Company,
by arrangement with Evans Brothers Limited, London, England.

Lerner Publications Company
A Division of Lerner Publishing Group
241 First Avenue North
Minneapolis, MN 55401

Website address: www.lernerbooks.com

Library of Congress Cataloging-in-Publication Data

Good, Keith.
    Shape it! : magnificent projects for molding materials / by Keith Good.
        p.   cm. — (Design it!)
    Includes index.
    Summary: Presents a range of projects giving readers hands-on
 experience with molding methods such as hand molding, stamping,
 and extruding, and encourages readers to design their own creations.
    ISBN 0-8225-3568-8 (lib. bdg. : alk. paper)
    1. Molding materials—Juvenile literature. [1. Molding materials.
 2. Handicraft] I. Title. II. Series.
 TS243.5.G66 1999
    731.4'3—DC21   99-36225

Printed in Hong Kong
Bound in the United States of America
1 2 3 4 5 6 -OS- 04 03 02 01 00 99

DESIGN IT!

SHAPE IT!

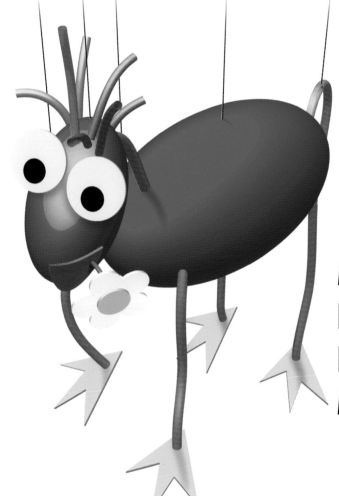

Magnificent Projects For Molding Materials

Keith Good

Lerner Publications Company • Minneapolis

About this book

About this series

This series involves young people in designing and making their own working technology projects, using readily available salvaged or cheap materials. Each project is based on a "recipe" that promotes success and stimulates the reader's own ideas. The "recipes" also provide a good introduction to important technology in everyday life. The projects can be developed to different levels of sophistication according to readers' ability and can reflect their other interests. The series teaches skills and knowledge in a fun way and encourages creative, innovative ideas.

About this book

One reason for young people to work with moldable materials is that they are a major part of their environment. They are literally surrounded by moldable materials as they live and work in brick and concrete homes and classrooms. They sit on molded chairs, cross molded bridges, make molded sandcastles, and eat molded bread, gelatin, lollipops, and cookies. The many plastic products that children use, such as toys and computer keyboards, are also molded. Not all moldable materials are suitable for children, but the ones in this book can be shaped quickly and easily so that plenty of ideas can be explored even when time is limited. Little equipment is needed. Fingers are often the best tools of all, giving real hands-on experience. Some processes like stamping (page 14) lend themselves to mass production, allowing this important concept to be discussed. Hollow forms (page 20) are among project parts that would be very difficult to make using other

methods and materials. Scientific ideas about the way materials behave (changing when heated, for example) can be brought to life through project work. Moldable materials are among the first that young people use. Through this book they can explore a range of materials and important processes and apply them creatively with increasing sophistication.

Safety

● Sharp tools are not necessary, but any that are chosen should be used under supervision.

● Using ovens requires adult supervision and thick oven mitts. Do not overheat plastic sheets.

● Do not use wallpaper paste containing fungicide to make papier mâché.

● Check that thread used to hold decorations around any part of the body breaks easily if snagged.

● Containers to be examined as part of designing molded containers should be well washed out before use.

● Ensure that good hygiene is observed when making food projects.

Contents

Introducing moldable materials
Making a display

Materials that can be molded play an important part in the world around us. They are very easy to shape into the things we need. When molds are used, many things such as house bricks, clay pots, and plastic toys can be copied quickly and cheaply. When shapes are molded, nothing is wasted by cutting away unwanted parts. Plasticine stays soft and can be molded over and over again. Other materials like concrete and plaster set hard when they are dry. Materials like clay are natural and very long-lasting – we can still see the footprints of dinosaurs where they molded clay by stepping in it!

You will need

- things molded from different materials
- pictures of molded things
- colored paper, pens, and other things for making a display

What to do

1. Set up a display area for a display about moldable materials. Use a table or other surface for standing things on and a wall or pinboard for pictures.

2. Cover your display area with cloth or colored paper so that your display will look good and things can be seen clearly.

3. Make your display as interesting and attractive as you can.

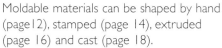

Moldable materials can be shaped by hand (page 12), stamped (page 14), extruded (page 16) and cast (page 18).

Getting ideas

When setting up your display, think of ways to show the different kinds of moldable materials and what they are used for. Try to show how important they are and give samples for your audience to handle. Look at how displays are done in your school, in stores, and in other places. Museums are a really good place to see how things can be displayed. What makes a good display? Some displays get people to do things as well as just look. Do activities help people to understand and make a display more fun? Could you design display activities? You could use a computer to find and print information, instructions, and labels.

Making moldable materials
From plaster to papier mâché

Lots of materials can be molded, but some need special equipment, and others are unsafe or expensive. Here are some materials you can use. Look out for these materials in use around you. Leave projects that need to dry in a warm place so that they will be ready quicker. Always protect your working area with a plastic sheet and wear an apron to protect your clothes.

You will need

To try out some of the projects in this book you will need:
- plain flour
- salt
- water
- air-drying clay
- plaster
- sandpaper
- Mod-roc
- newspaper
- tissue paper
- foam plastic sheet (such as Plastazote or Formafoam)

Salt dough

1. Mix 2 measures of plain flour and 1 measure of salt. Gradually add 1 measure of water. Knead (squeeze, mix, and squash) the dough until it molds well. Wire hooks and other objects can be molded into the dough before baking.

2. When you have made your project, put it on a baking sheet in an oven to harden. **Safety:** Get adult help when using the oven. Cook at a low heat (about 200°). A long time on a low heat is best. Big or thick pieces will take longest to harden.

3. Let your project cool before painting it.

Air-drying clay

1. Mold the clay to the shape you want using fingers, plastic picnic knives and forks, and other objects like old pen caps that you can find. Beads can be made by molding around a rod.

2. Leave to dry well. Keep unused clay well wrapped in plastic to keep it soft.

Normal clay has to be fired (baked) in an oven. Bricks and tiles are made this way.

Plaster

1. Mix 3 measures of plaster with 1 of water. Stir well. Tap the mold gently as you pour in the liquid.

2. Take out of the mold when dry. Smooth sharp edges with sandpaper. Paint when very dry. Wire, hooks and other things can be embedded in the plaster before it sets.

This process is called casting (see page 18). Plaster casts are used to decorate walls and ceilings and to make lasting copies of animal tracks and footprints.

Foam plastic sheet

Only use special foam sold for molding and modeling. Brand names include Plastazote and Formafoam.

1. **Safety**: Get adult help to warm the sheet in an electric oven set to 300°. A 6mm-thick sheet will need to be warmed for about one minute. A smelly or sticky sheet is too hot.

2. Use oven mitts to take the sheet out of the oven. When it has cooled a little but is still warm, mold it to the shape you want. For example, a warm strip can be wound around a pencil to make a spiral.

3. Sheets can also be cut with scissors, sewn, stapled, or glued. Plastic foam is used for gym mats, shoe liners, and swimming pool floats. (This kind of foam is NOT suitable for projects.) Look out for other uses.

Mod-roc

This is a special cloth with plaster that you drape over a form (like a plasticine shape or a plastic bottle). It is like the plaster used to hold broken limbs still while they get better.

1. Coat the form with petroleum jelly.

2. Take enough Mod-roc to cover the form twice and cut it into squares.

3. Dip each square into tepid water for four or five seconds. When the bubbles stop, squeeze the Mod-roc gently and smooth it over the form to dry.

Papier mâché

1. Tear pieces of newspaper into strips.

2. Make a paste by adding water to flour a little at a time.

3. Put the strips in a tray and brush them with the paste.

4. Drape the pasted strips over a form to make the shape you want. Use tissue paper for the last layer to make a smooth surface.

Balloons and bottles can be used as forms. Strips can be pressed into a mold (such as a dish).

5. Leave to dry well before painting.

Papier mâché is French for "chewed newspaper." It has been used to make furniture, trays, bowls, and other household items.

Moldable materials to eat
Basic recipes for cookies and breads

Safety

Food safety and hygiene is important!
- Tie back long hair and wear a clean apron.
- Wash hands with soap, dry with a clean towel.
- All equipment must be clean and only used for food. Wash after use and dry with a clean cloth.
- Work on a clean surface.
- Don't cough or sneeze over food or equipment.
- Remember that some people must not eat certain foods.
- Get adult help with food safety and using an oven.

You will need

Basic cookie recipe
- 1/2 cup soft margarine
- 1/2 cup brown or white sugar
- 2 cups plain flour
- 1 egg
- pinch of salt

Cookie recipe

1. Beat the margarine and the sugar together in a bowl.

2. Beat the egg and add it to the mixture.

3. Add the flour and salt and mix to make dough.

4. Roll out the dough and cut or stamp out the shapes you want.

5. Put the shapes on a greased baking sheet and bake at 375° for about 15 minutes.

You will need

Basic bread recipe
- 4 cups flour
- 1 cup warm water
- 1 level teaspoon salt
- 2 tablespoons sunflower oil
- 1 packet dried yeast

Bread recipe

1. Combine the ingredients in a large bowl.

2. Add spices if desired, and mix well. (Divide the recipe if you want to try different combinations.)

3. Knead the dough.

4. Let the dough rise for at least 20 minutes.

5. Put on a baking sheet and bake in an oven set to 425° for 15-20 minutes.

Drawing
Modeling your ideas on paper

Drawings are an important way of giving information. Instruction books, car repair manuals, dressmaking patterns, and all sorts of plans are used every day. They give information quickly and can be understood by people who don't speak the same language. Drawings are also a good way for people to explore their own ideas. Drawing helps you to think things through – keeping all your ideas in your head is difficult! Although molded materials are quick to shape, drawing can be an even quicker way to try out ideas. Drawing also helps you to talk to others about your ideas and gives you a record of the thinking that led to your final design.

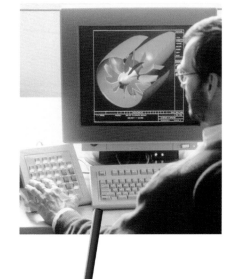

You will need

- 8.5" by 11" paper
- drawing and coloring equipment

Tip

Keep your eyes on the cross. Don't stare at the pencil point.

Concentrate on the last line you did.

1

2

What to do

1. Draw a small cross and practice drawing straight lines to it. Start close to the cross and then gradually draw from farther away.

2. Draw a line and practice drawing lines parallel to it. Start with lines close together and gradually draw them farther apart.

3. Attach your paper over the square grid on page 28 with paper clips. Use paper that allows the grid to show through. Paper used for photocopying is good for this. Draw faint frames to help you draw lots of squares and circles. Next, try designing interesting counters for a board game. The *center line* (*C/L*) is important when the object is *symmetrical* (both halves are the same).

You can use the square grid to help you draw projects. To help you cut out the clay or dough to the right shape, trace your drawing to make a *pattern* or *template*.

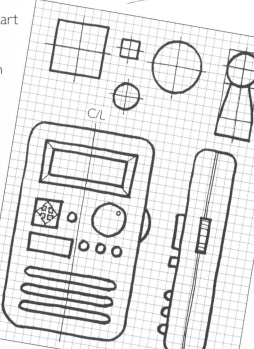

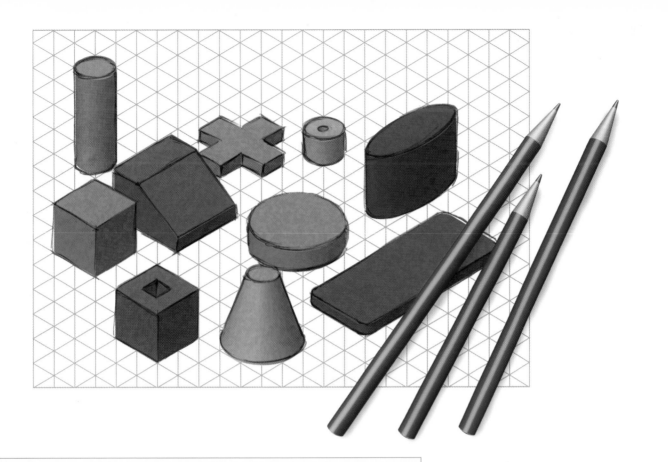

4. Attach your paper over the *isometric* grid on page 29 and practice drawing the shapes shown above. Draw faint frames to help you. Many things you need to draw can be made up of simple shapes like these.

5. Using the same grid, make a drawing putting the shapes together to make an interesting object. You could design a product of the future or one of the other projects in this book.

6. Lots of moldable materials projects have rounded edges and corners. Practice by drawing shapes lightly with sharp edges, then round them off.

7. You can make your drawings look more real by shading them to show the light and dark surfaces. Top surfaces are often the lightest. If you are using colored pencils, use darker and lighter shades.

You can also use lines, dots, and perhaps color to show what something is made from.

Molding by hand
Pots, cord pulls, and zipper pulls

Hard materials can only be shaped with a tool. Moldable materials are soft enough to be shaped just with hands. Potters and sculptors often shape clay with their hands. Cooks mold food by hand. Many things we use need to be shaped so that they fit our hands well. Squeezing plasticine is one way to find a good shape for a handle, which can then be copied in a harder material. You can shape air-drying clay and salt dough just using your hands. When these materials harden the things you make can do useful jobs. Try the cord and zipper pull activity on the next page, but first explore what your hands can do with moldable materials.

You will need

- plasticine
- salt dough (see page 7)
- air-drying clay (see page 7)
- white glue
- thin garden wire
- paints

What to do

1. Take some plasticine and notice how it feels and behaves when you squeeze it, stretch it, twist it, and press it.

2. Try to make a small thumb pot. Can you make a shape that will float?

3. Roll the plasticine. Twist and coil it.

Have you made any shapes that could be used for something – especially if they were made of a material that set hard?

Reinforced concrete has steel rods inside to add strength. Shaping the dough or clay around wire will make your molding stronger, especially where it is very narrow.

When dough and clay are hard they are quite brittle, so don't make shapes that are very thin or narrow. Wire is also useful when you want to connect something to your molding.

reinforced concrete with steel rods

steel rod

What to do

1. Read "Getting ideas" below and think about what your project will be for and what it will be like. Draw some different ideas.

2. For a cord pull, mold around a rod so that there will be a hole for the cord. For a zipper pull, make a wire shape. A bent shape will grip the dough or clay better than a straight one.

3. Mold the shape you want. Leave air-drying clay to dry and harden. With adult help, bake a salt dough project and let it cool.

4. Paint your project and leave it to dry. Varnish with watered-down white glue, which will dry clear.

You could tie your project to a zipper pull, but a small metal split ring is a neater way to join the two together. Get an adult to help you attach the ring. Tie a knot in the cord to keep a cord pull in place.

Getting ideas

Look at cord pulls, zipper pulls, and other handles, especially ones that are pulled. Where are they found? What are they made from? How do they feel? Draw some that you find and make notes about them. What do you like and dislike? How could you improve them?

Play with a piece of dough to find out what shapes you can make. What patterns and textures can be made by pressing things into the dough? Remember to avoid thin shapes that might break.

As well as looking good, your pull will need to fit the fingers or hand. Think about how it will be used. Will the user be wearing gloves? Where will your pull be used? Consider sports bags, clothing, roller blinds, light switches, and other ideas.

Rolling, stamping, and marking

Making shapes and decorating them

Rolling is used to shape metal and other materials. The hot metal is squeezed between sets of rollers that are closer and closer together until the metal is the right thickness. Rolling pins are used in kitchens to roll pastry into thin sheets. Clay and other moldable materials can also be rolled. *Stamping* is a quick way of cutting out shapes from flat sheets. Coins, cookies, milk bottle caps, and other things are stamped out in large numbers. Cooks use cutters to cut out shapes from pastry. Stamping or pressing into material is a good way to decorate surfaces and lots of marks can be made the same. You can roll, stamp, and mark to make puzzles, games, and parts for other projects like the one on page 19.

You will need

- plasticine (for trying out your ideas first)
- air-drying clay
- two strips of wood
- rolling pin kept for clay work, or short piece of broom handle
- interesting things to use for stamping and marking

What to do

1. On a smooth board, roll out a slab of clay large enough for your project. Using two strips of wood will give you an even thickness. Thin strips can be taped together if a thicker slab is needed.

2. A paper template will help if you want to cut out shapes that are the same. Fold and cut out a template if you want to make symmetrical shapes (with both halves the same shape).

3. Use a modeling tool or plastic knife to cut around your template. You can also cut out pieces by measuring and marking on the slab. Use a wet finger to make any rounded edges that you want.

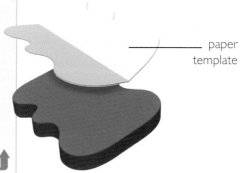

paper template

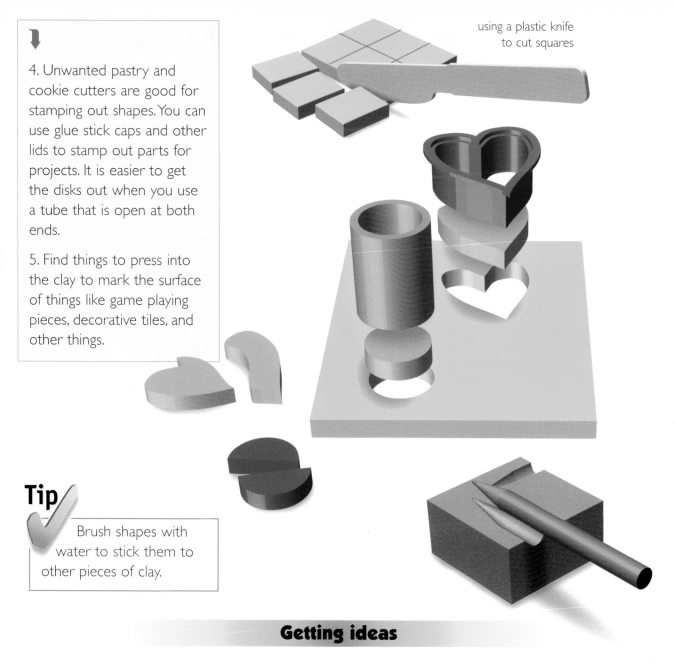

using a plastic knife to cut squares

4. Unwanted pastry and cookie cutters are good for stamping out shapes. You can use glue stick caps and other lids to stamp out parts for projects. It is easier to get the disks out when you use a tube that is open at both ends.

5. Find things to press into the clay to mark the surface of things like game playing pieces, decorative tiles, and other things.

Tip

Brush shapes with water to stick them to other pieces of clay.

Getting ideas

You could design jigsaw puzzles by rolling, stamping out, and marking a design or picture, then cutting it into pieces. Parts could be made to fit together. You could try out your design in cardboard or plasticine first. Don't make it too easy or too hard. You could design games that use playing pieces that have been stamped out and marked. The playing board could be made in cardboard or even in fabric. You could design your own dice. You could use a computer to write rules and make numbers and shapes for your game board. If you don't have a color printer you could color a black-and-white printout. Could you design a game that could also be played by people with sight problems or with eyes covered? Which are easiest to feel – pieces stuck on or marks pressed into the surface? Finding out about games from different countries could help you have your own ideas.

15

Extruding moldable materials
Making a minibeast

Extruding is a way of shaping material by pushing it through a hole so that it comes out the same shape as the hole. Toothpaste coming out of a tube is being extruded. Cakes are decorated by squeezing icing through a shaped nozzle. Some cookies and other foods are extruded. When they are hot enough, metals like aluminum are soft enough to be extruded to make window frames and other products. Plastic pipes, insulation on electric wires, and other everyday products are also made by extrusion. You can extrude materials like salt dough, air-drying clay, and plasticine by pushing them through sieves, graters, and other objects with holes in them. One use for the shapes is to make new creatures or minibeasts that you have designed.

You will need

- plasticine
- salt dough or air-drying clay
- an unwanted tea strainer, sieve and other things with holes to try
- nail brush

What to do

1. Try out extruding by pushing some plasticine through a sieve or tea strainer. This will make shapes that look like hair or fur that can be scraped off using a plastic knife.

2. Salt dough and air-drying clay can also be pushed through holes to make lasting projects. Experiment with other tools such as a garlic press.

3. Different amounts of water in the dough or clay will change how it behaves. You could divide your clay or dough into same-sized pieces and add a different number of drops of water to each. Add a few drops of water at a time and work it in well before deciding whether to add more.

Very wet clay pushes straight through a grater well, but when you want to take shavings off a block, is a drier block better?

A toy "dough factory" will extrude lots of different shapes if your material is soft enough. Wash it well before the clay or dough dries.

Creating a creature

1. Use books and CD ROMs to find out about small creatures. Use the ideas on this page to help you design a new creature.

2. You can use extruding with processes like hand-molding (see page 12), rolling, stamping, and marking (see page 14).

3. Thin parts made from clay or dough are very fragile, so mold in wire to make parts like legs and feelers. Thin parts of a body can also be made stronger with wire.

4. Add any extruded or stamped out parts. Brush with water to help pieces stick. You could mark the surface with toothpicks and other tools.

5. Get adult help to bake a salt dough creature (see page 7) and let it cool. Let an air-drying clay creature dry well. Paint carefully, then stick on more parts if you wish.

Getting ideas

To help you get ideas for your new creature or minibeast, take a close look at pictures of real ones.

What do they look like? How do they protect themselves? What do they eat and how do they find their food? How will *your* minibeast live? What will it eat? Are there animals that hunt your creature? Does it hunt other creatures, or live on plants, candle wax, or litter? Find out about food chains in nature. What senses will your creature have? How will it find or catch its food? How will it protect itself? How will it move?

What will your minibeast be called? Could it be a robot bug? It could have an everyday name, a scientific name, and a nickname. Design a care manual to tell people about your creature and how to look after it. Use a computer to do this if you can.

Casting
Making models with molds

Pouring molten metal, liquid plaster, wet concrete, gelatin, or some other material into a mold is called *casting*. When the material sets or cools, it hardens and becomes the same shape as the mold. Casting is a quick way of making copies of something, and very little material is wasted. Candles, ice cubes, paving stones, and parts for buildings and bridges are all made by casting. You can design your own electronic product using casting. Many of the electronic objects around us started as models. Designers use non-working models to try out their new ideas and show them to others. Your product does not really have to work, so use your imagination!

You will need

- plaster
- water
- molded plastic packaging or clear plastic bubble wrap
- plasticine

Tip

Two castings from the same mold can be glued together. Smooth the back of the castings with sandpaper.

What to do

1. Find some packaging to use as a mold. Choose something that is the right shape for your design ideas. Use plasticine to hold the mold level if you need to.

2. Mix up the plaster (see page 7) and pour it into the mold, tapping the mold to help it flow. Let it set, then take the casting out.

3. When the casting is completely dry, use sandpaper to take off any sharp edges. The project can now be painted. Stickers and parts that you find can be stuck on to make controls, screens, and other things.

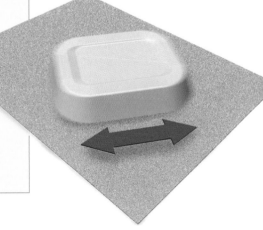

What to do

Here is how to make another kind of mold:

1. Make a slab of plasticine (see page 14) on a board. Press a strip of plasticine firmly around the edge of the slab to make the wall of the mold.

2. Use modeling tools and other objects to press in and shape the mold. You can also add plasticine shapes.

Remember that you are working in reverse. Make a dent in the plasticine and it will come out as a bump on your cast. Add a raised shape to the slab and it will come out as a dip.

3. Pour in the plaster and let it set.

4. Take the plaster cast out of the mold and paint it. When dry, varnish with watered-down white glue.

Tip

Put your mold (on its board) in a large bowl or a tray in case it leaks.

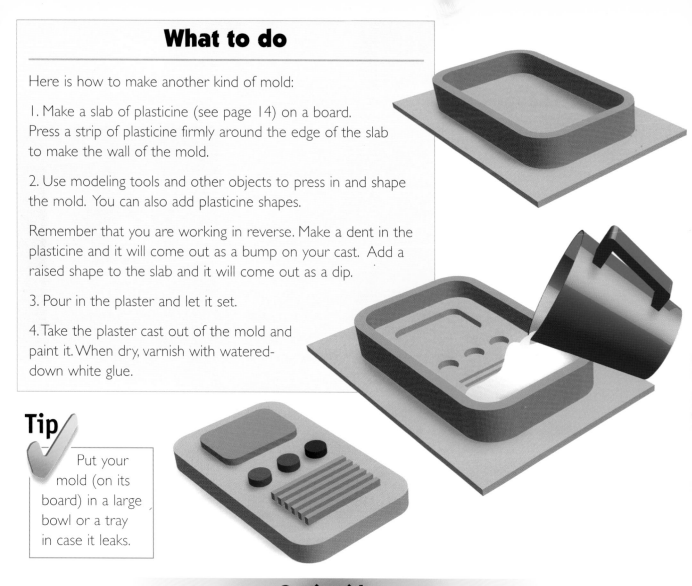

Getting ideas

Think about designing and modeling an electronic project of the future. Electronic products are a very important part of the world around us. How many can you list? Imagine if there were no calculators, computer games, pagers, hand-held computers, or mobile phones. Look at products that already exist in your home, school, stores, and catalogs. How could they be improved? What extra things could they do? What new products would *you* like to have? What will people really need in the future? What would the *real* version of your product do? Ask other people for their ideas. A busy parent might like a device to keep a toddler from wandering off. (It would need to be safe and kind to the child!) Think about what controls you would need and where they should be put. Think about color, shape, size, appearance, and safety. Write and draw a user's manual or instruction leaflet for your product. For a very realistic manual, use a computer (word-processing, desktop publishing, graphics, color printing).

19

Hollow molding
Making puppets for a play

Hollow moldings weigh very little for their size because they are full of air! Large things can be made without using much material because the moldings are hollow. Plastic footballs, dolls, and bottles are all hollow moldings. Some are made by blowing softened plastic into a mold. Other things, such as canoes, are made by joining two moldings together. You can make hollow moldings using papier mâché or Mod-roc (see page 8). You could make some shapes in both materials to compare them. The spaces inside could be used to hold things like batteries, wires, bulbs, and other working parts. You could use a hollow molding when designing a puppet!

You will need

- papier mâché or Mod-roc (see page 8)
- plasticine
- petroleum jelly
- balloons

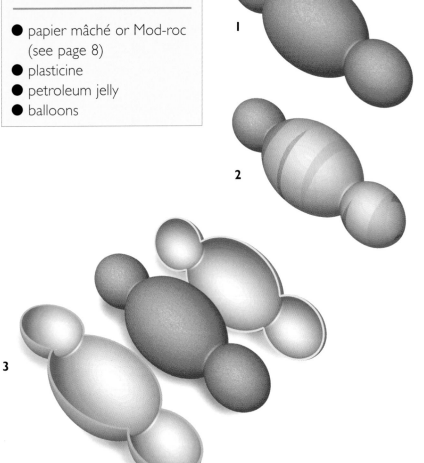

What to do

1. Decide what shape you want your puppet to be and mold the shape in plasticine. Coat it with petroleum jelly. This will keep the covering from sticking to it.

2. Choose your covering – papier mâché or Mod-roc. Smooth several layers over the plasticine. Put in a warm place to dry well.

3. Get adult help to cut the molding in half and take out the plasticine. This would be a good time to make holes ready for tying on things like legs, neck, and control strings (see page 21).

4. Use strips of the covering to join the halves together again.

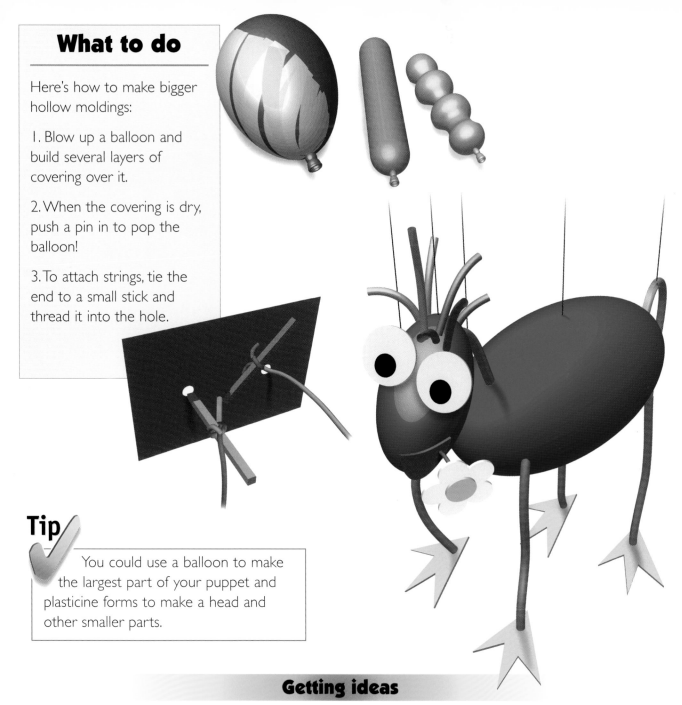

What to do

Here's how to make bigger hollow moldings:

1. Blow up a balloon and build several layers of covering over it.

2. When the covering is dry, push a pin in to pop the balloon!

3. To attach strings, tie the end to a small stick and thread it into the hole.

Tip

You could use a balloon to make the largest part of your puppet and plasticine forms to make a head and other smaller parts.

Getting ideas

There are many kinds of puppets. What will your puppet be like? What sort of character will it be? It could just hang from one string or piece of elastic, or have several moving parts. You could get together with friends and write a play. Perhaps your puppet or play could be used to teach young children something important. If you can, use a computer to word-process a neat script.

Save your script on disk so that you can easily change and improve it. You could go on to plan a performance and design a simple puppet theater. Hanging your puppet up when it is not being used will keep the strings from tangling. Puppets have been made for a very long time and in many different places. Use a CD ROM and books to find out more about the world of puppets.

21

Molded containers

Designing practical packages

Among the molded products you see every day are many different bottles and containers. Shower gels, shampoos, deodorants, perfumes, aftershaves, makeup, and toothpaste all come in molded packages. Designers often use models to try out their ideas and to avoid expensive mistakes. Containers don't just have to hold a product. They usually have to make people want to buy it, and give information. They should also be easy to use. Designers need to think about all the energy and materials that go into containers and what should happen to them when they are empty. How many molded glass and plastic containers are there in your home? You don't have the machines to make plastic and glass containers, but you can design and model them.

Safety: Some liquids used in the home can be harmful. Get adult advice and make sure that containers you look at or take parts from are washed out thoroughly first.

You will need

- plasticine
- air-drying clay or salt dough (page 7)
- paper
- coloring equipment
- bottle caps

What to do

1. Think about the design of your container and what would go in it if it was real. Read these two pages and draw different ideas before you start making anything (see page 10).

2. Start by making a basic shape and mold your container from that. A few examples are shown here. You could try out your ideas in plasticine first because it won't dry out while you experiment.

3. To make your project look more real you could add a lid, cap, or other part that you have found. Make a shape to hold it in place when you glue it. ↑

4. Use fingers, modeling tools, or a plastic knife to make the shapes you want. Cut-out shapes can be stuck on after brushing them with water.

5. Let your model dry if you used air-drying clay. Get adult help to bake a salt dough model and let it cool.

6. Paint your model. To make it shiny, brush it with a mixture of water and white glue.

7. Design a label for the front of your container. You could design one for the back as well. Look at real labels to get ideas.

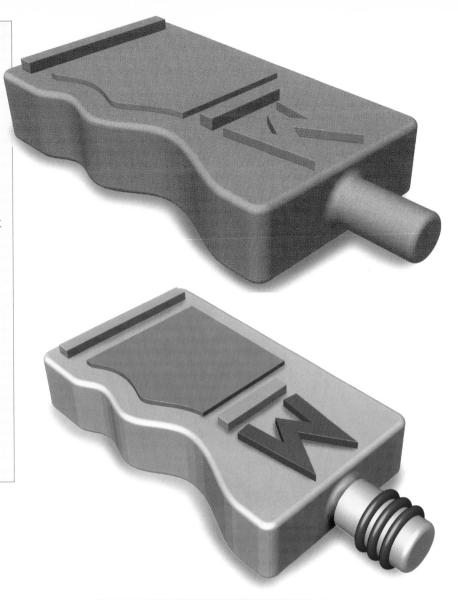

Getting ideas

What would your container hold? It could be something to use in the shower, bathroom, or kitchen, or somewhere else. Could it hold a special potion or magic medicine? Perhaps it will be home for a genie! Do people have hobbies that give you ideas? Look at containers in your home as a starting point – but don't just copy them. What do you like about them and what do you dislike? Are they easy to use? Do they fit your hand? Think about how your container might be used. Will it be opened and used with wet or soapy hands? Will it stand on a shelf or hang on a hook? Could we re-use or recycle more containers? Think of new uses for used containers. Could yours be designed to have other uses?

Look at labels. Yours could have information, warnings, bar codes, a list of ingredients, and pictures. How will your labels and container make people want to buy the product? If you can, use a computer to make realistic labels.

Decorating people with moldings

Making jewelry

People have been wearing decorations for thousands of years. Some decorations tell us things, perhaps that someone is married. Other decorations show that the wearer has certain beliefs or that they belong to a club or other group. People have used almost every material to decorate themselves, including wood, feathers, animal teeth, and shells. Jewels, gold, silver, and other metals are often used as decoration. Metals are sometimes melted and poured into molds to make jewelry. This is called casting (see page 18).

You can shape some moldable materials to make decorations. Remember that decoration has to be practical too. Make sure that yours are not too heavy or fragile. It is very important that decorations are safe and comfortable as well as beautiful.

Safety

Use thin thread to make sure anything hung around the neck or other part of the body *breaks easily* if it snags on something.

You will need

- papier mâché pulp (page 8)
- drinking straws
- air-drying clay (page 7)
- foam plastic sheet (page 8)
- paints and thread
  - other materials that you want to add

Medals, badges, brooches, and pendants

1. Cut out shapes in clay or dough. Stick on any shapes with water and do any molding. Add a safety pin or make a hole for hanging.

2. Paint your projects when the dough is baked or the clay is dry. Varnish with watered-down white glue if you want a shiny finish.

Beads

1. Roll out a slab of salt dough or air-drying clay and cut or stamp out the shapes you want. Make holes with a toothpick for threading.

2. To make beads the same size, cut equal pieces from a cylinder. Mold the shapes and make holes. Lightweight beads can also be made by flattening some papier mâché pulp and folding it around a piece of a straw.

Bracelets and bangles

1. Overlap and tape a circle of thin cardboard so that it will slip onto your wrist.

2. Use a plastic bottle full of water to hold the circle while you add layers of papier mâché pulp. You could also try making a Mod-roc bangle (see page 8).

You could stick on extra shapes and paint when dry.

Bangles, bracelets, and other decorations can be molded from foam plastic sheet (see page 8).

1. Get adult help to warm the sheet in an electric oven.

2. Mold the sheet around something that is the right size and shape for your project.

Spirals can be made by winding strips of foam around a pencil or stick.

Combine different colors by pressing pieces together under a board when they come out of the oven.

You can press things into the warm sheet to emboss it.

Getting ideas

Your decoration could be for a special theme party. You could design a matching set of fun jewelry. You might need to use colors that go with clothes or a costume. Look out for feathers, silk flowers, sequins, shells, small bells, and other things to use with your moldable materials. Could you add things to make a rattling or some other noise as the wearer moves? Beads can be worn around different parts of the body or sewn or tied onto fabric. Try cutting up extruded shapes (see page 16) to make beads. Could you invent an award and make a medal for it? You could use a computer to make a certificate too. You could design a decoration that plays a part in a story or play. It might be very valuable or have special powers. Choose a lightweight material like papier mâché for large decorations. Look at other pages in this book for more methods you could use. For very large pieces look at page 20. To help with your own designs, find out about decorations and jewelry in different parts of the world and throughout history.

Molding more things to eat
Cookies and bread with a difference

Food products such as chocolate, pastry, and marzipan are moldable. Popsicles and fruit gelatin are cast in molds. Special cakes can be beautifully decorated by extruding icing through different-shaped nozzles. Cookies are fun foods that can be made from moldable materials. Bread dough is also easy to mold into many different shapes. Bread is an important food in many different cultures. Take a close look at different breads like whole wheat, herb, crusty, pita, and rye. If you can get some to try, describe what they smell and taste like. Look in bakeries, supermarkets, and books to find out about new and traditional cookies and breads. **Important**: Anything that will touch food must be really clean. Read about food safety and hygiene on page 9. Get adult help when using a hot oven.

You will need

- basic cookie mix (page 9)
- 1 cup powdered sugar
- small candies, dried fruit, other edible decorations
- basic bread mix (page 9)
- herbs, seeds, and spices
- toothpicks, colored paper, other things for decoration

Cookies

1. Design your cookies. Read these pages for help.

2. Use the basic mix or add flavoring to it. Ideas for different flavors include grated lemon or orange peel (wash the fruit first), grated chocolate, dried fruit, and other safe flavorings. Ask adult advice and do not add too much. **Remember** that some people are allergic to nuts and other foods.

3. Cut or stamp out the shapes you want. Cutting gives you more choice. Try out shapes on paper and make a template. This is especially useful when you want lots of shapes the same.

4. Bake the cookies (see page 9).

5. If you want to add icing, mix 1 cup of powdered sugar with 1 tablespoon of warm water.

6. Spread the icing with a knife and press in candies and other decorations. You can also draw and write on the cookies using an icing syringe.

Bread

1. Design your bread project. Read these pages for help.

2. Use the basic recipe to make dough. Before adding water, mix in herbs and spices if you want them.

3. Make the shapes you want. Dough can be molded and stuck together with water. Use a garlic press to extrude hair shapes (see page 16). Seeds and other things can be sprinkled on. Brushing your design with egg white will make it shiny when baked.

4. Bake (see page 9) and let it cool well before tasting. Hot bread can burn!

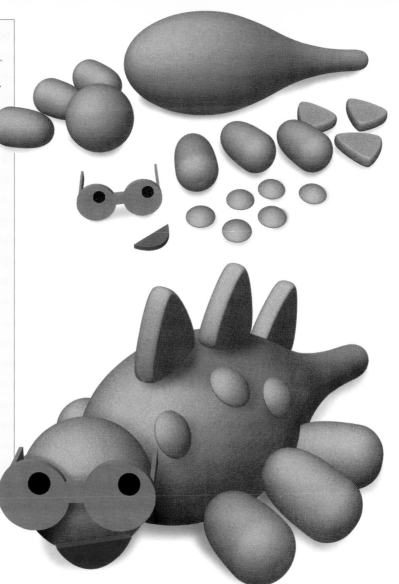

Getting ideas

How would you decorate your cookies to show what flavor they are? Make up interesting names for them. You could make cookies into a gift for someone.

Design a package that will protect the cookies and look attractive. You could think about bags, boxes, packets, tubes, and other ideas. Think of different ways to decorate your package and tell people about what is inside. Perhaps you could use computer graphics and word-processing. Collect pieces of

ribbon and other attractive things for decoration. Collect and look at different cookie packages.

Think about ways to flavor your bread. Will it be for a special party or picnic? Decorations could be added, like toothpicks and paper sails to make bread boats. Use a computer to print paper decorations that tell people about your bread or sandwich. You could design sandwiches with unusual but tasty fillings.

Grid for drawing

Square grid (see page 10)

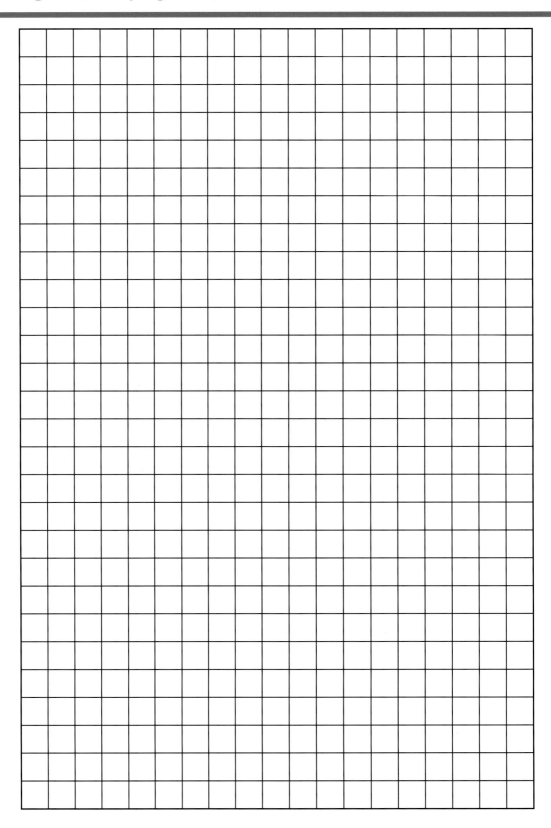

Top of grid

Index

COOKIES
AND
CANDIES

KIDS' KITCHEN

COOKIES AND CANDIES

WICKEDLY DELICIOUS RECIPES
FOR JUNIOR CHEFS

NICOLA FOWLER

CHARTWELL
BOOKS, INC.

A Quintet Book

Published by Chartwell Books
A Division of Book Sales Inc.
PO Box 7100
Edison, New Jersey 08818-7100

This edition produced for sale in the
U.S.A., its territories and dependencies
only.

ISBN 0-7858-0386-6

This book was designed and produced by
Quintet Publishing Limited
6 Blundell Street
London N7 9BH

Creative Director: Richard Dewing
Designer: Ian Hunt
Project Editor: Diana Steedman
Editor: Emma Tolkien
Photographer: Jeremy Thomas

Typeset in Great Britain by
Central Southern Typesetters, Eastbourne
Manufactured in Singapore by
Eray Scan Pte Ltd
Printed in Singapore by
Star Standard Industries (Pte) Ltd

ACKNOWLEDGEMENTS

Special thanks to Spencer and Victoria
Dewing, Tom Lolobo, Gee Hyun Kim,
Nick Seruwagi, James and Lucy Stuart, and
to Bob McNiff of the Burlington Junior
School, New Malden, Surrey.

PUBLISHER'S NOTE

Children should take great care when
cooking. Certain techniques such as
slicing and chopping or using the stove,
oven, or broiler can be dangerous, and
extreme care must be exercised at all
times. Adults should always supervise
while children work in the kitchen.

As far as methods and techniques
mentioned in this book are concerned, all
statements, information, and advice given
here are believed to be true and accurate.
However, the author, copyright holder,
nor the publisher can accept any legal
liability for errors or omissions.

Contents

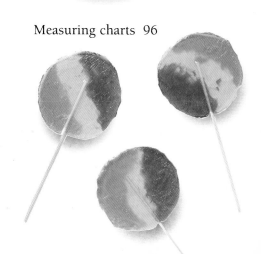

Introduction

Cooking for the family and friends is lots of fun and *Cookies and Candies* will show you just how easy it can be, too. It is never too soon to start learning the simple basics of food preparation and cooking. So, if this is your first attempt, start with the easy recipes first and, when you become more confident, move on to the more involved recipes.

You will find many old favorites together, with a selection of unusual and creative ideas. The recipes are easy to follow, with clear step-by-step pictures showing the different techniques required. For safety reasons, some of the photographs are bordered in red, and the instructions are highlighted in bold text. Read Before You Begin first – it is full of important hints and tips – then put on your apron, and have fun and enjoy what you are doing.

Happy Cooking

Before You Begin

Here are a few simple rules you should always follow whenever you are cooking.

▌ Never cook unless there is an adult there to help you. Certain tasks such as chopping with sharp knives, heating liquids, and using electrical equipment can be dangerous if not done properly and should always be supervised by an adult. We have highlighted these tasks throughout the book so that you know when to ask for help.

▌ Always read through the recipe before you start and gather together all the ingredients and equipment you will need. Weigh and prepare the ingredients as instructed by the recipe.

▌ It is not unusual to find cookery books with both metric and standard units. Always stick to one system as they are not interchangeable. A useful chart for temperatures, weights and measures is on page 96.

▌ Hygiene is important in the kitchen. Tie your hair back if it is long, wash your hands before touching food, and wear an apron to keep your clothes clean.

▌ Keep your working area clean and neat. Clean up as you go along, and have a sponge handy to wipe up spills as soon as they happen.

▌ Be careful not to burn yourself when cooking. Wear oven gloves when handling hot pans and baking sheets, and never put your fingers into hot mixtures. If you do accidentally burn yourself, hold the burn under cold water and tell an adult immediately.

▌ Never use electrical equipment near water.

▌ Never leave gas and electric burners on if you have to go out of the kitchen, and remember to turn everything off when you have finished.

▌ Finally, always leave the kitchen work surfaces neat and clean.

Some Cooking Techniques

▌ **Knives** must always be used with great care and with an adult present. Never hold the food you are cutting with fingers outstretched; instead, hold the food with your fingertips tucked under and clear of the knife edge. Do not rush and concentrate on what you are doing.

▌ Always **preheat the oven** so that it is at the correct temperature throughout the cooking time.

▌ **Greasing and lining pans** prevents cookies and candies from sticking. To grease, put a little vegetable oil onto the pan and spread with a brush or a piece of paper towel, over the base and sides. If the recipe tells you to line the pan, then grease the pan first, place a piece of parchment paper in the base, and then grease the paper.

SIFTING FLOUR

▌ **Sifting** flour and powdered sugar gets rid of any lumps and helps to make the mixture light and airy.

▌ **Rubbing fat into flour** with your fingertips is a method used in some cookie recipes and in pastry. It is important to keep your fingers cool and to lift the mixture well to add lots of air. It is ready when it looks like bread crumbs.

▌ To **separate an egg**, crack it firmly against the rim of a small bowl and allow the white to fall into the bowl. You may need to transfer the yolk carefully between the two halves of shell once or twice to allow all the white to fall out. Put the yolk into a separate bowl.

▌ When **adding eggs** to a creamed butter and sugar mixture, beat the eggs lightly first, add a little at a time, and beat in well after each addition to prevent the mixture from curdling.

ROLLING AND KNEADING DOUGH

▌ **Roll dough** on a floured surface with a floured rolling pin to prevent it from sticking. Add more flour if necessary.

▌ To **knead** dough, place it on a lightly floured surface and work it between your hands until it is smooth.

Folding means gently adding ingredients to a mixture which contains a lot of air, for example beaten egg whites. Use a metal spoon and do not beat the mixture.

Melting chocolate in a bowl over a pan of hot water prevents the chocolate from overheating and cooking (rather than just melting). Make sure the bowl is big enough to fit over the top of the pan and that the water does not touch the bowl. Or use a double boiler.

Piping icing and cookie dough is easy with a bit of practice, and using a piping bag enables you to control the pressure with your fingers. Experiment with nozzles in different sizes and shapes.

Safety in the Kitchen

Heat Use oven gloves when handling anything that is hot. Never try to take something out of a hot oven without using oven gloves.

▌ When using saucepans, hot dishes, or baking sheets, use both hands and check first that there is nothing in the way of where you are going to place them once you have removed them from the stove or the oven.

▌ Place hot pans and dishes on a trivet or board. If you feel that a pan or dish is too heavy for you to handle, ask someone to do it for you.

▌ When pans are on the stove, keep the handles turned away from you. Angle them to the side and not over the burners which, if turned on, will heat the handles, making them too hot to handle.

▌ Do not overfill saucepans as they will then be too heavy to lift. There is also the danger that the contents may boil over if the pan is filled close to the brim.

▌ Call an adult immediately if a fire breaks out. Do NOT try to deal with it yourself.

▌ Do not be tempted to test for heat by placing your hand or fingers on or in anything that may be hot.

▌ Turn off the stove, microwave, and any electrical implements that have been used as soon as you have finished with them.

▌ Always informing an adult when you are beginning to cook; then you must also tell them when you have finished so they can check that electrical implements are safely turned off.

CHAPTER ONE

Cookie Know-how

Oat Crunch Cookies

Makes about 18 cookies

YOU WILL NEED

1 stick unsalted butter,
  cut into pieces
2 level tablespoons corn syrup
⅓ cup light brown
  soft sugar
⅝ cup all-purpose four, sifted
½ cup oats
1 egg, lightly beaten

1 **Preheat the oven to 300°F**. Lightly grease two baking sheets.

2 Put the butter and syrup into a medium saucepan and **melt together over a gentle heat, stirring occasionally. Remove the pan from the heat** and place on a heatproof surface.

3 Add all the dry ingredients to the saucepan. Finally add the egg.

4 Stir the mixture with a wooden spoon until it is well combined.

5 Using two teaspoons, place heaped teaspoons of the mixture on the greased baking sheets. Leave plenty of room between each cookie to allow them to spread during cooking.

6 Press the cookies down with the back of a fork to flatten them slightly.

7 **Cook in the preheated oven for about 15 minutes** until they are golden brown. Wearing oven gloves, **remove the baking sheets from the oven** and leave the cookies to cool for five minutes. Then, using a spatula, transfer to a cooling rack and leave until cold.

Shortbread Shells

Makes 12 cookies

YOU WILL NEED

2 sticks unsalted butter,
   at room temperature
½ cup sugar
   (plus extra for sprinkling)
1¼ cups all-purpose flour, sifted
¼ cup ground rice
pinch salt

1 **Preheat the oven to 400°F**. In a large bowl, **beat the butter and sugar together with an electric mixer** until the mixture is creamy and light.

2 With a wooden spoon, beat in the remaining ingredients until the mixture is well combined.

3 Turn the dough onto a surface dusted with flour and knead lightly until it is smooth.

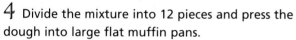

4 Divide the mixture into 12 pieces and press the dough into large flat muffin pans.

5 **Bake in the preheated oven for 12 to 15 minutes** until lightly golden. Wearing oven gloves, **remove the pan from the oven** and leave the cookies to cool for 10 minutes, then carefully turn out onto a cooling rack. When completely cold, sprinkle with a little sugar.

Fruity Oat Squares

Makes 9 cookies

YOU WILL NEED

¾ cup dried, ready-to-eat apricots

¾ cup dried, ready-to-eat prunes

1 stick unsalted butter, cut into pieces

¼ cup light brown soft sugar

3 tablespoons honey

1 cup oats

large pinch salt

1 **Preheat the oven to 350°F.** Grease a 7-inch square shallow pan. Then **using a small knife, cut apricots and prunes into small pieces** and set aside.

2 Put the butter, sugar, and honey into a large saucepan and **melt over a gentle heat, stirring occasionally.**

3 **Remove the pan from the heat** and place on a heatproof surface. Add the chopped apricots and prunes, oats, and salt to the saucepan.

4 Stir the mixture until well combined.

5 Spoon the mixture into the greased pan and, using the back of a large spoon, press the mixture down firmly until the surface is level.

6 **Bake in the preheated oven for 25 minutes.** Wearing oven gloves, **remove the pan from the oven** and leave to cool on a wire rack for 5 minutes. Then **mark into 9 squares with a small, sharp knife** and leave the cookies in the pan until completely cold.

7 When cold, loosen the squares and remove the pan.

Nice Spice Cookies

Makes about 25 cookies

YOU WILL NEED

1 small lemon, skin scrubbed
  clean
1½ sticks unsalted butter, at
  room temperature
⅓ cup light brown soft
  sugar
1 egg, lightly beaten
¾ cup all-purpose flour, sifted
2 teaspoons ground cinnamon
½ teaspoon ground nutmeg
½ teaspoon ground ginger

1 **Preheat the oven to 350°F.** Lightly grease two baking sheets. Finely grate the rind of the lemon and set aside.

2 In a large bowl, **beat the butter and sugar with an electric mixer** until the mixture is creamy and light.

CHEF'S TIP

☞ Squeeze the juice of the lemon; add a little hot water and sugar. Cool the syrup and then top up with water for a refreshing drink.

3 Then add the egg, a little at a time, **beating well between each addition**. Add the lemon rind.

4 Add the sifted flour and spices and, using a wooden spoon, stir well until the mixture forms a stiff dough.

5 Knead the dough lightly in the bowl and, using your hands, shape the dough into approximately 25 small balls.

6 Place the balls on the greased baking sheets, leaving room between each for spreading during cooking.

7 Make a pattern on the cookies by flattening each slightly with the prongs of a fork dipped in flour in two directions to make a crisscross pattern. **Bake in the preheated oven for about 10 minutes until golden**. Wearing oven gloves, **remove the baking sheets from the oven** and cool on a heatproof surface for five minutes. Then, with a spatula, transfer the cookies to a cooling rack and leave until cold.

Jeweled Sugar Cookies

Makes about 25 cookies

YOU WILL NEED

½ stick unsalted butter,
  at room temperature
½ cup sugar
1 egg plus 1 egg yolk, lightly
  beaten
¾ cup all-purpose flour, sifted
½ teaspoon baking powder
pinch salt
a few drops almond extract
4 tablespoons multicolored
  sugar crystals

1 In a large bowl, **beat the butter and sugar with an electric mixer** until the mixture is creamy and light.

2 Add the egg, a little at a time, **beating well between each addition**.

3 Add the flour, baking powder, salt, and almond extract, and mix with a wooden spoon until well combined.

4 Turn the dough onto a surface dusted with flour and knead lightly until it is smooth. Then wrap the dough in foil or in a plastic bag and refrigerate for 30 minutes.

5 **Preheat the oven to 350°F.** Lightly grease two baking sheets. Sprinkle a surface with flour and roll out the dough with a floured rolling pin until it is about ¼ inch thick.

6 Using a heart-shaped cutter, cut cookies out of the dough, lightly re-kneading and re-rolling the dough until it is all used up. Place the cookies on the baking sheets, leaving a little space around each one.

7 Sprinkle the cookies with colored sugar crystals. **Bake in the preheated oven for about 10 minutes until pale golden.** Wearing oven gloves, **remove the baking sheets from the oven**, place on a heatproof surface and leave the cookies to cool for 5 minutes. Then, with a spatula, transfer the cookies to a cooling rack and leave until cold.

Chocolate Crunch Bars

Makes about 16 bars

YOU WILL NEED

½ pound graham crackers

1 x 4-ounce chocolate
  honeycomb bar

1 stick unsalted butter, cut into
  pieces

4 tablespoons corn syrup

1 cup semisweet chocolate chips

⅛ cup powdered sugar

1 Line the base and sides of a 9-inch pie plate with a piece of aluminum foil.

2 Put the graham crackers and chocolate honeycomb bar into plastic bags and crush coarsely with a rolling pin.

3 Put the butter and syrup into a large saucepan and **melt over a gentle heat, stirring. Remove the pan from the heat** and place on a heatproof surface.

4 Stir the chocolate chips into the melted butter and syrup until the chocolate has melted and the mixture is smooth.

5 Add the crumbs and chocolate honeycomb to the melted mixture in the saucepan and stir well to combine.

6 Pour the mixture into the prepared pie plate and smooth the surface with the back of a spoon. Let it cool, and then refrigerate for at least 2 hours.

7 Turn the chilled mixture onto a chopping board, remove the foil, and sprinkle with powdered sugar. **Cut into slices to serve.**

Coconut Crisp

Makes about 18 cookies

YOU WILL NEED
⅛ cup all-purpose flour
½ stick unsalted butter,
   at room temperature
½ cup sugar
2 egg whites
⅛ cup dessicated coconut

1 **Preheat the oven to 350°F.**
Lightly grease two large baking
sheets. Sift the flour into a bowl.

2 In a separate bowl, **beat the
butter and sugar with an
electric mixer** until the mixture
is creamy and light.

3 Add the egg whites to the
butter and sugar and **gently
mix until just combined.**

4 Finally, add the sifted flour and dessicated coconut and mix with a spoon.

5 Put heaped teaspoons of the mixture onto the prepared baking sheets, leaving plenty of room for spreading.

6 **Bake in the preheated oven for about 8 minutes** until the cookies are brown around the edges.

7 Wearing oven gloves, **remove the baking sheets from the oven**, place on a heatproof surface, and leave the cookies to cool on the baking sheets for 5 minutes. Then, with a spatula, transfer the cookies to a wire rack to cool completely.

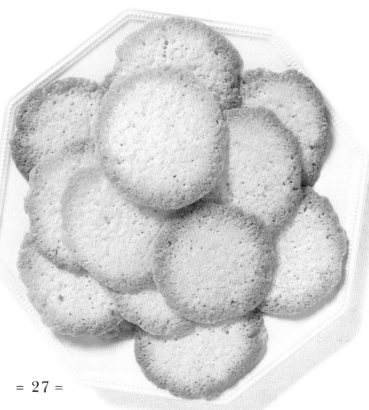

Fruit 'n' Nut Brownies

Makes about 16 brownies

YOU WILL NEED

3 ounces baker's chocolate, broken into small pieces

¾ stick butter, at room temperature

¾ cup sugar

3 eggs, lightly beaten

1 teaspoon vanilla extract

⅓ cup self-rising flour, sifted

pinch salt

½ cup walnuts, chopped

½ cup raisins

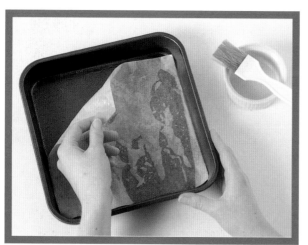

1 **Preheat the oven to 350°F.** Grease, line base, and grease base of a shallow 8-inch square backing pan.

2 Half fill a medium-sized saucepan with water and **bring to a boil. Take the saucepan off the burner** and place on a heatproof surface. Place the chocolate in a bowl that will sit on top of the saucepan without touching the water. Put the bowl on top of the pan and stir the chocolate occasionally until it has melted and is smooth.

3 Meanwhile, put the butter and sugar into a separate large bowl and **beat with an electric mixer** until the mixture is creamy and light. Add the egg, a little at a time, **beating well between each addition**.

4 Add all the remaining ingredients, except the melted chocolate, and stir well to combine.

5 Finally, add the melted chocolate and stir until well mixed.

6 Spoon the mixture into the prepared tin and **bake in the preheated oven for approximately 25 minutes**.

7 Wearing oven gloves, **remove the pan from the oven** and place on a cooling rack. Leave the brownies in the pan to cool and then cut into squares.

CHEF'S TIP

☞ Brownies should be soft and chewy in the middle, so be careful not to let them overcook and become dry.

Blueberry Bites

Makes about 20 cookies

YOU WILL NEED

1½ sticks unsalted butter,
  at room temperature
½ cup sugar
1 egg, lightly beaten
1 teaspoon vanilla extract
¾ cup self-rising flour, sifted
½ cup dried blueberries

1 In a large bowl, **beat the butter and sugar with an electric mixer** until the mixture is creamy and light.

2 Add the egg, a little at a time, **beating well between each addition.**

3

4

5

3 Add the vanilla extract and then the flour, mixing with a spoon until well combined.

4 Stir in the blueberries.

5 Turn the dough onto a surface dusted with flour and knead lightly until it is smooth. Form the dough into a roll, wrap, and refrigerate for 30 minutes.

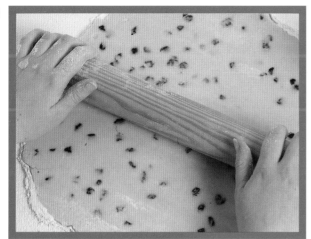

6 **Preheat the oven to 350°F.** Lightly grease two baking sheets. Sprinkle a little flour on a surface and roll out the chilled dough with a floured rolling pin until it is about ¼ inch thick.

7 Cut out cookies using a crinkle-edged round cookie cutter. Lightly re-knead and re-roll the dough until it is all used up. Place the cookies on the prepared baking sheet and **bake in the preheated oven for about 15 minutes until golden-brown.** Wearing oven gloves, **remove the baking sheets from the oven**, place on a heatproof surface, and leave the cookies to cool for 5 minutes. Then, with a spatula, transfer the cookies to a cooling rack, and leave to cool completely.

Peanut Butter Cookies

Makes about 16 cookies

YOU WILL NEED

½ stick unsalted butter, at room
   temperature
½ cup crunchy peanut butter
½ cup light brown soft sugar
1 egg, lightly beaten
1 teaspoon vanilla extract
¾ cup all-purpose flour, sifted
1 teaspoon baking powder
pinch of salt
16 unsalted and blanched
   peanuts

1 **Preheat the oven to 350°F.**
Grease two baking sheets. In a
large bowl, **cream together the
butter and peanut butter with
an electric mixer.**

2 Add the sugar and continue
to **beat until the mixture is
creamy and light.**

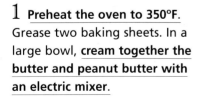

3 Add the egg, a little at a time, **beating well between each addition.**

4 Now add all the remaining ingredients, except the peanuts, mix well with a spoon, and form into a ball.

5 Turn the dough onto a surface dusted with flour and knead lightly until it is smooth. Sprinkle the surface with flour again and roll out the dough with a floured rolling pin until it is about ¼ inch thick.

6 Cut the dough into circles using a plain round cutter. Lightly re-knead and re-roll dough until it is all used up.

7 Put the cookies on the prepared baking sheets and place a peanut in the center of each cookie. **Bake in the pre-heated oven for about 10 minutes until golden brown.** Wearing oven gloves, **remove the baking sheets from the oven**, place on a heatproof surface, and leave the cookies to cool for 5 minutes. Then, with a spatula, transfer the cookies to a cooling rack and leave until cold.

Giant Double Choc Chip Cookies

Makes 8 cookies

YOU WILL NEED

2 sticks unsalted butter,
   at room temperature
1 cup sugar
1 egg, lightly beaten
1 cup self-rising flour, sifted
⅛ cup cocoa powder, sifted
pinch salt
½ cup white chocolate drops

1 **Preheat the oven to 350°F.**
Lightly grease two large baking
sheets. In a large bowl, **beat the
butter and sugar with an
electric mixer** until the mixture
is creamy and light.

2 Add the egg, a little at a
time, **beating well between
each addition.**

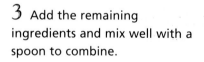

3 Add the remaining ingredients and mix well with a spoon to combine.

4 Turn the dough onto a surface dusted with flour and knead lightly until it is smooth.

5 Form the dough into eight equal balls.

6 Lightly flour the surface again, and roll each ball into a flat cookie with a floured rolling pin until each cookie is about 5 inches in diameter.

7 Place the cookies on the prepared baking sheets and **bake in the preheated oven for about 12 minutes.**

8 Wearing oven gloves, **remove the baking sheets from the oven**, place on a heatproof surface, and leave the cookies to cool for 10 minutes. Then, with a spatula, transfer the cookies to a wire rack to cool completely.

CHEF'S TIP

☞ You could make smaller cookies by following the recipe to Step 4, then rolling the dough and cutting with a cookie cutter. Smaller cookies may need a little less cooking time.

Fast Freezer Cookies

Makes about 25 cookies

YOU WILL NEED

1½ sticks butter, at room
   temperature
1 cup sugar
1 egg, lightly beaten
¼ cup ground almonds
⅝ cup self-rising flour, sifted
pinch salt

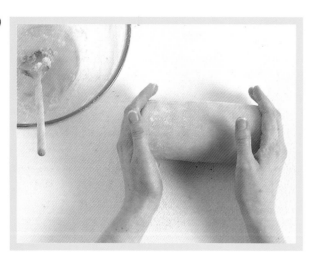

1 In a large bowl, **beat the butter and sugar with an electric mixer** until the mixture is creamy and light. Add the egg, a little at a time, **beating well between each addition**.

2 Add all the remaining ingredients and beat well with a spoon.

3 Shape the dough into a roll. It will be very soft.

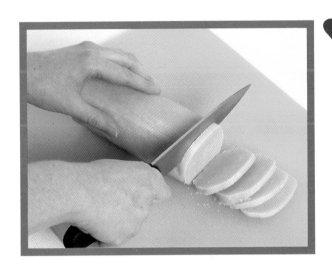

4 Wrap the dough roll in waxed paper and put into the freezer until you want to bake the cookies.

5 Take the dough out of the freezer 10 minutes before you want to bake the cookies. **Preheat the oven to 400°F** and grease a baking sheet. **Slice the required number of cookies – about ¼ inch thick – from the roll using a sharp knife.**

6 Place the cookies on the greased baking sheet. Do not worry if they are a bit curly and do not lie flat on the sheet. They will flatten out in the oven.

7 **Bake in the preheated oven for about 9 minutes** until golden brown. Wearing oven gloves, **remove the baking sheet from the oven**, place on a heatproof surface, and leave the cookies to cool on the sheet for 5 minutes. With a spatula, transfer to a wire rack to cool completely.

CHAPTER TWO
Cookie Crazy

Animal Farm Gingerbread

Makes about 35 small cookies

YOU WILL NEED

1 stick unsalted butter, cut into pieces
½ cup light brown soft sugar
1 tablespoon corn syrup
¾ cup all-purpose flour
1 teaspoon baking powder
2 teaspoons ground ginger
2 cups powdered sugar, sifted
8–12 teaspoons hot water
pink and yellow food coloring

1 Put the butter, sugar, and syrup into a large saucepan and **melt over a gentle heat, stirring. Remove the pan from the heat**, place on a heatproof surface, and allow the mixture to cool for 10 minutes.

2 Sift the flour, baking powder, and ginger together and add to the saucepan. Mix well and form the dough into a ball. Wrap the dough and refrigerate for 30 minutes.

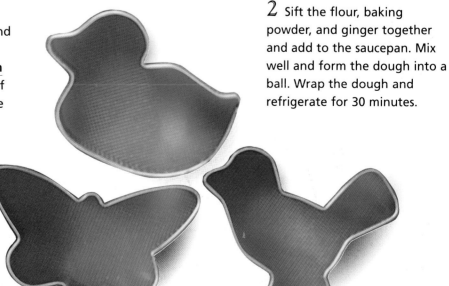

it is all used up. **Bake in the preheated oven for 7–8 minutes until golden brown.** Wearing oven gloves, **remove the baking sheets from the oven**, place on a heatproof surface, and leave the cookies to cool for 5 minutes. Then, with a spatula, transfer the cookies to a rack and leave until cold.

5 Meanwhile, put the powdered sugar in a bowl. Add about half the water and mix well. Then continue adding more water, a teaspoon at a time, stirring well after each addition, until you achieve a fairly thick icing that will hold its shape when piped.

3 **Preheat the oven to 350°F.** Lightly grease two baking sheets. On a lightly floured surface, roll out the chilled dough with a floured rolling pin until it is about ¼ inch thick.

4 Using animal-shaped cookie cutters, cut shapes from the dough and place on the prepared baking sheets. Lightly re-knead and re-roll the dough until

6 Divide the icing up into small bowls, leave one white and color each of the others a different color with a few drops of food coloring, say pink for pigs, yellow for chicks.

7 Using one color at a time, put the icing into a piping bag fitted with a plain writing nozzle. Cover the remaining bowls of icing with a damp cloth until you are ready to use them. Pipe faces and patterns onto the cookies and leave them to dry before serving.

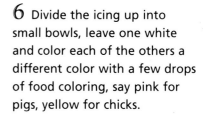

CHEF'S TIP

☞ You can buy a variety of animal-shaped cookie cutters from supermarkets and department stores.

Cookie Crumbles

Makes about 16 cookies

YOU WILL NEED

1¼ sticks unsalted butter,
at room temperature

¼ cup sugar

¾ cup all-purpose flour, sifted

⅛ cup cornstarch

⅓ cup jelly

¼ cup light brown soft sugar

1 **Preheat the oven to 350°F.**
Grease two baking sheets. Put
aside ¼ stick of the butter. Put
the remaining butter in a large
bowl and **beat with an electric
mixer until soft**. Add the sugar
and **beat again** until the
mixture is creamy and light.

2 Put aside ¼ cup of the flour.
Add the remaining flour and
cornstarch to the butter and
sugar mixture, mix well with a
spoon, and form the dough into
a ball.

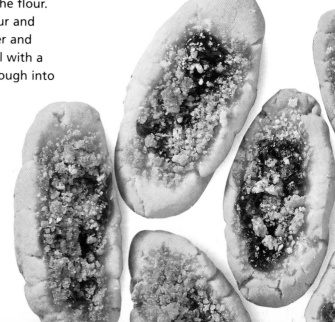

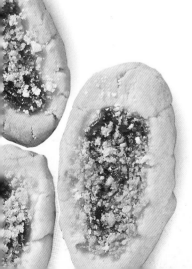

3 Take small pieces of the dough and roll into "sausages." Place each on the baking sheets, allowing room to spread during cooking. With your finger, make a small indent down the center of each cookie.

4 Fill the indentation with a little jelly.

5 Now make the crumble topping. Put the reserved butter and flour into a bowl and rub together with your fingers until the mixture looks like bread crumbs. Stir in the brown sugar.

6 Sprinkle the crumble topping over the cookies. **Bake in the preheated oven for about 15 minutes.** Wearing oven gloves, **remove the baking sheets from the oven**, place on

a heatproof surface, and leave the cookies to cool for 10 minutes. With a spatula, transfer the cookies to a cooling rack and leave until cold.

CHEF'S TIP

☛ This is a good recipe for using up little bits of jelly left in the bottom of the jar. Try different flavors in each cookie.

Ice Cream Sandwiches

Makes about 18

YOU WILL NEED

1 stick unsalted butter,
  at room temperature
¾ cup sugar
1 egg, lightly beaten
¾ cup self-rising flour, sifted
1 pint ice cream, any flavor

1 In a large bowl, **beat the butter and sugar with an electric mixer** until the mixture is creamy and light. Add the egg, a little at a time, **beating well** between each addition.

2 Add the sifted flour, mix well with a spoon, and form the dough into a ball. Cover and refrigerate for about 30 minutes.

3 **Preheat the oven to 350°F.** Lightly grease two baking sheets. On a floured surface, knead the chilled dough lightly and roll out with a floured rolling pin until the dough measures about 12 inches square.

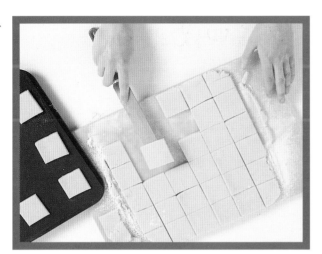

4 **Using a sharp knife, cut the dough into strips at 2-inch intervals.** Then cut across the strips at 2-inch intervals so that you have 36 squares. With a palette knife, lift the squares and place them on the prepared baking sheets, leaving room for spreading.

5 **Bake in the preheated oven for about 10 minutes until golden brown.** Wearing oven gloves, **remove the baking sheets from the oven**, place on a heatproof surface, and leave the cookies to cool for 10 minutes. Then, with a spatula, transfer the cookies to a wire rack to cool completely.

6 When the cookies are cold, remove the ice cream from the freezer and allow to stand for five minutes at room temperature. Then spoon ice cream onto half of the cookies, **pushing it down with a knife** to give a flat surface. Place a cookie on top of the ice cream to make a "sandwich."

Twinkle Stars

Makes about 20 cookies

YOU WILL NEED

1½ sticks unsalted butter,
   at room temperature
¾ cup self-rising flour, sifted
¼ cup light brown soft sugar
½ teaspoon vanilla extract
pinch salt
edible silver sugar balls
edible silver powder

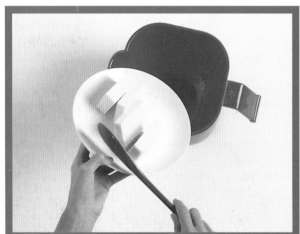

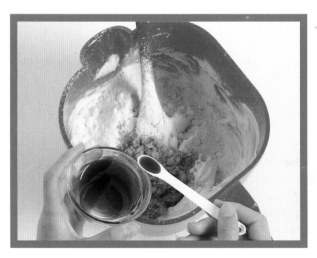

1 **Preheat the oven to 375°F.** Lightly grease two baking sheets. Place the butter in a bowl and beat with a wooden spoon until soft.

2 Sift the flour into the bowl, then add the sugar, vanilla extract, and salt. Mix well. The dough will be crumbly, so roll it into a ball with your hands. Transfer to a lightly floured surface. Knead lightly until smooth.

3 Roll out the dough with a floured rolling pin until it is about ¼ inch thick.

5 Press a silver sugar ball onto each point of the stars. **Bake in the preheated oven for 13–15 minutes until the cookies are golden brown.** Wearing oven gloves, **remove the baking sheets from the oven**, place on a heatproof surface, and leave thecookies to cool for 5 minutes. Then, with a spatula, transfer the cookies to a cooling rack and leave until cold.

6 Use a basting brush to paint the outline of the cookies carefully with a thin coat of silver dust.

4 Cut out cookies using a star-shaped cookie cutter. Using a spatula, transfer to a baking sheet. Lightly re-knead and re-roll dough until it is all used up.

Alpha-Bitties

Makes about 20 letters

YOU WILL NEED

1 stick unsalted butter,
  cut into pieces
¾ cup all-purpose flour, sifted
½ cup sugar
1 tablespoon poppy seeds
1 egg, lightly beaten
few drops of vanilla extract
1½ cups powdered sugar, sifted
6–8 teaspoons hot water
⅛ cup nonpariels

1 **Pre-heat the oven to 400°F.** Lightly grease two baking sheets. Put the butter and flour into a large bowl and rub the butter into the flour with your fingertips until the mixture looks like bread crumbs.

2 Stir in the sugar and poppy seeds, followed by the egg and vanilla extract.

3 Form the dough into a ball, turn onto a lightly floured surface, and knead the dough lightly.

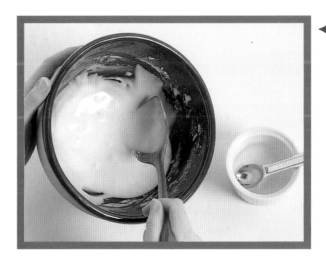

minutes. Then, with a spatula, transfer the cookies to a wire rack and leave until cold. Meanwhile, put the powdered sugar into a bowl, add about half the water, and mix well. Then continue adding more water, a teaspoon at a time, stirring well after each addition, until you achieve a thick icing that will spread.

6 With a small palette knife, spread the icing on the cookies. If necessary, dip the palette knife in hot water from time to time to make the spreading easier. While the icing is still wet, sprinkle the cookies with nonpariels and leave to dry.

4 Sprinkle the surface with flour again and roll out the dough with a floured rolling pin until it is about ¼ inch thick. Using letter-shaped cookie cutters or cardboard templates and a sharp knife, **cut out letters from the rolled dough**.

5 Put the cookies carefully onto the prepared baking sheets and **bake for about 10 minutes until golden brown**. Wearing oven gloves, **remove the baking sheets from the oven**, place on a heatproof surface, and leave the cookies to cool for 5

Wiggly Worms

Makes about 10 cookies

YOU WILL NEED

1 stick unsalted butter,
   at room temperature
½ cup powdered sugar, sifted
1 egg, lightly beaten
⅝ cup all-purpose flour, sifted
1½ cups powdered sugar, sifted
6–8 teaspoons hot water
few drops of pink food coloring

1 **Preheat the oven to 350°F.** Line a baking sheet with parchment paper.

2 In a large bowl, **beat the butter and ½ cup of the powdered sugar with an electric mixer** until the mixture is creamy and light. Add the egg, a little at a time, beating well between each addition.

3 Add the flour and mix well with a spoon.

4 Spoon the cookie mixture into a large piping bag fitted with a ½-inch plain nozzle.

5 Pipe about 10 wiggly shapes onto the parchment paper, leaving space for spreading.

6 **Bake in the preheated oven for approximately 10 minutes** until golden brown. Wearing oven gloves, **remove the baking sheet from the oven**, place on a heatproof surface, and leave the cookies to cool for 5 minutes. Then, with a spatula, transfer the cookies to a cooling rack and leave until cold.

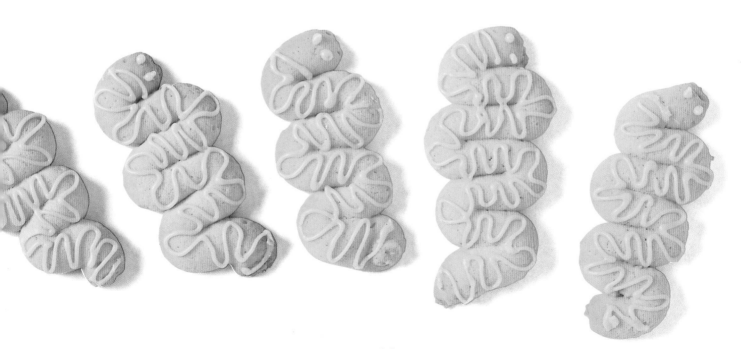

7 Meanwhile, make the icing. Put the remaining powdered sugar in a bowl, add about half the water, and mix well. Then continue adding more water, a teaspoon at a time, stirring well after each addition, until you achieve a thick icing that will hold its shape when piped. Add a few drops of food coloring and put the icing in a small piping bag fitted with a plain nozzle. Pipe eyes and a zigzag pattern onto each "worm." Leave to dry before serving.

CHEF'S TIP

Place a piping bag in a small bowl or wide glass and fold the sides over the edge to support the bag while you fill it.

Christmas Cookies

Makes about 35 cookies

YOU WILL NEED

½ stick unsalted butter,
  at room temperature
½ cup light brown soft sugar
1 egg plus 1 egg yolk, lightly
  beaten
¾ cup all-purpose flour, sifted
½ teaspoon baking powder
pinch salt
2 teaspoons mixed spice
1 cup powdered sugar, sifted
8–12 teaspoons hot water
yellow, green, and red food
  coloring
narrow colored ribbon or string

1 In a large bowl, **beat the butter and sugar with an electric mixer** until the mixture is creamy and light. Add the egg, a little at a time, **beating well** between each addition.

2 Add the flour, baking powder, salt, and mixed spice and mix with a wooden spoon until well combined. Turn the dough onto a surface dusted with flour and knead lightly until it is smooth. Then wrap the dough in foil or in a plastic bag and refrigerate for 30 minutes.

3

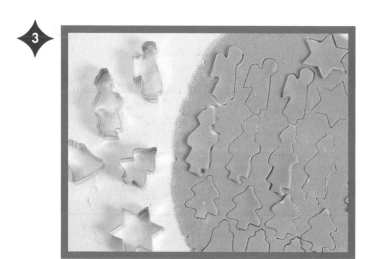

4

5

3 **Preheat the oven to 350°F**. Lightly grease two baking sheets. Sprinkle a surface with flour and roll out the dough with a floured rolling pin until it is about ¼ inch thick. Using Christmas cookie cutters, cut cookies out of the dough, lightly re-kneading and re-rolling the dough until it is all used up.

4 With a toothpick, make a little hole at the top of each cookie. Place them on the baking sheets, leaving a little space around each one.

5 **Bake in the preheated oven for about 10 minutes** until golden brown. Wearing oven gloves, **remove the baking sheets from the oven**, place on a heatproof surface, and leave the cookies to cool for 5 minutes. Then, with a spatula, transfer the cookies to a cooling rack and leave until cold.

6 Meanwhile, put the powdered sugar into a bowl, add about half the water, and mix well. Then continue adding more water, a teaspoon at a time, stirring well after each addition, until you achieve a thick icing that will hold its shape when piped. Divide the icing up into small bowls, leave one white, and color each of the others a different color with a few drops of food colouring, say yellow, green, and red. Using one color at a time, put the icing into a piping bag fitted with a plain nozzle. Cover the remaining bowls of icing with a damp cloth until you are ready to use them. Pipe outlines and patterns onto the cookies.

7 When the icing has dried, thread a piece of colored ribbon through the hole in each cookie, tie in a bow, and hang the cookies on the Christmas tree.

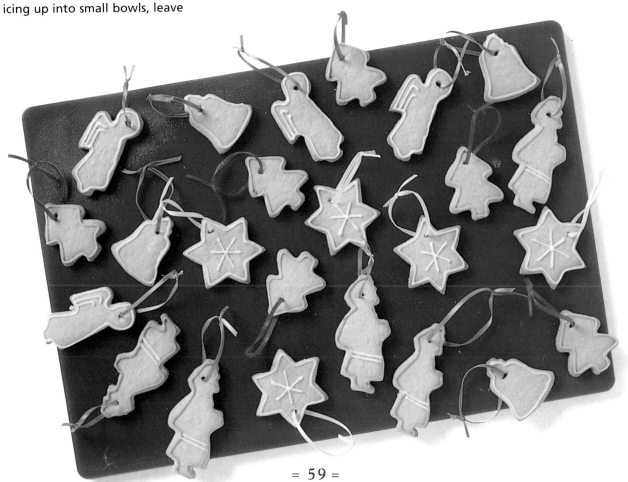

Crazy Faces

Makes about 20 cookies

YOU WILL NEED

½ stick unsalted butter,
   at room temperature
½ cup sugar
1 egg plus 1 egg yolk, lightly
   beaten
3/4 cup all-purpose flour, sifted
½ teaspoon baking powder
pinch salt
½ teaspoon vanilla extract
ready-to-roll icing in assorted
   colors

1 **Preheat the oven to 350°F.** Lightly grease two baking sheets. In a large bowl, **beat the butter and sugar with an electric mixer** until the mixture is creamy and light. Add the egg, a little at a time, **beating well between each addition**. Add the flour, baking powder, salt, and vanilla extract, and stir with a spoon until well combined.

2 Turn the dough onto a surface dusted with flour and knead lightly until it is smooth. Sprinkle the surface with flour again and roll out the dough with a floured rolling pin until it is about ¼ inch thick. Cut out cookies with a plain round cookie cutter and place them on the prepared baking sheets. Lightly re-knead and re-roll the dough until it is all used up.

3

4

5

6

3 **Bake in the preheated oven for about 10 minutes** until they are lightly browned. Wearing oven gloves, **remove the baking sheets from the oven**, place on a heatproof surface, and leave the cookies to cool for 5 minutes. Then, with a spatula,

transfer the cookies to a cooling rack and leave until cold.

4 Meanwhile, lightly dust a surface with powdered sugar and roll out each color of the ready-to-roll icing.

5 Using cookie cutters, aspic cutters, and/or a small knife, **cut out shapes** from the ready-to-roll icing to make faces.

6 Dampen the underside of the icing slightly with water to stick the icing to the cookies.

Butterfly Whirls

Makes about 10 cookies

YOU WILL NEED

1½ sticks unsalted butter,
   at room temperature
¾ cup self-rising flour, sifted
¼ cup light brown soft sugar
½ teaspoon vanilla extract
pinch of salt
green food coloring
5 licorice strands

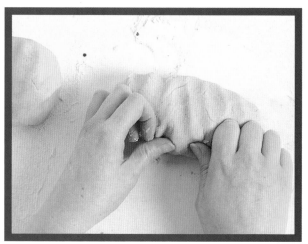

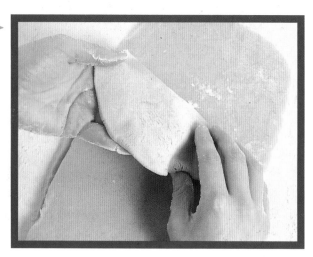

1 **Preheat the oven to 375°F.** Lightly grease two baking sheets. Put the butter in a bowl and beat with a wooden spoon until soft. Add the flour, sugar, vanilla extract, and salt and mix well. Form the dough into a ball and knead on a lightly floured surface until smooth. Divide the dough in two.

2 Add a few drops of green food coloring to one half of the dough and knead well until the color is evenly mixed in.

3 On a floured surface, roll each piece of dough with a floured rolling pin until each is approximately 5 x 10 inches. Place one piece of dough on top of the other.

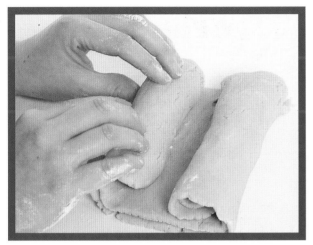

4 From one short end, roll up the dough to the middle. Do the same from the other end.

5 **With a knife, slice through the roll at ½-inch intervals** and place the cookies on the prepared baking sheets, leaving room for spreading. Press each cookie lightly with your hand to flatten slightly.

6 **Cut the licorice into 3-inch lengths,** fold in half, and press one into the middle of each cookie, overlapping the edge, to make the antennae. **Bake in the preheated oven for approximately 15 minutes** until golden brown. Wearing oven gloves, **remove from the oven,** place on a heatproof surface, and leave the cookies to cool for 5 minutes. Transfer to a cooling rack and leave until cold.

Meringue Mushrooms

Makes about 20

YOU WILL NEED

2 eggs
½ teaspoon white vinegar
½ cup sugar
¼ stick unsalted butter
¼ pound baker's chocolate,
  broken into small pieces
1 tablespoon milk

and put the whites into a large, very clean bowl. Add the vinegar to the egg whites and **beat with an electric mixer** until the foam becomes very stiff.

2 Add about half the sugar and **continue beating**. Then add the rest of the sugar and **beat until the mixture is thick and glossy** and will stand in stiff peaks.

3 Spoon the meringue into a piping bag fitted with a ½-inch plain nozzle. Pipe about 20 large "mushroom caps" approximately 1½ inches in diameter onto one of the prepared baking sheets, giving them plenty of space to spread. Then pipe about 20 smaller "blobs" onto the other prepared baking sheet for the "stalks."

1 **Preheat the oven to 275°F**. Line two baking sheets with baking parchment. Separate the eggs

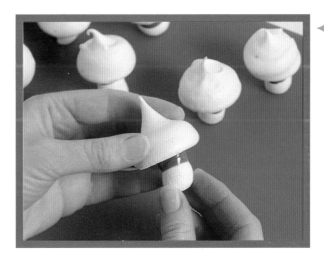

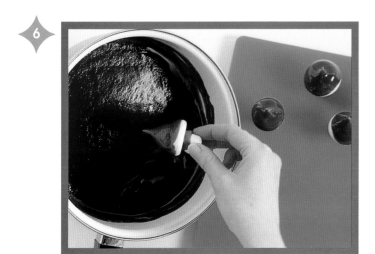

4 **Bake in the preheated oven** for about one hour until very lightly browned. Wearing oven gloves, **remove the baking sheets from the oven**, place on a heatproof surface, and transfer the meringues to a cooling rack with a palette knife. Meanwhile, **melt the butter in a small saucepan, remove from the heat,** and place the pan on a heatproof surface. Add the chocolate to the pan and stir until the chocolate has melted completely.

5 **With a knife, level off the top of the meringue "stalks,"** dip each in melted chocolate and attach a stalk to the base of each "cap." Leave in the refrigerator until cold and completely set.

6 **Heat the milk until hot but not boiling** and add it to the chocolate remaining in the saucepan. Stir until the mixture is smooth and well mixed. Dip each mushroom cap into the chocolate mixture to coat. You must do this very carefully to make sure you do not pull the stalks off the caps. Stand the mushrooms on their stalks and leave to dry.

CHAPTER THREE
Sweet Treats

Marzipan Fruit

Makes ½ pound fruit

YOU WILL NEED

½ pound white prepared
marzipan
red, yellow, brown, and green
food coloring

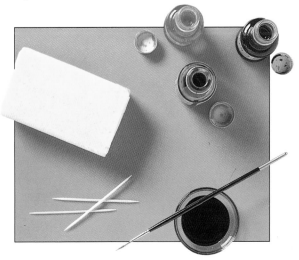

1 On a surface dusted lightly with powdered sugar, knead the marzipan well to soften it.

2 Divide the marzipan into four balls.

3 Taking one ball of marzipan at a time, add a few drops of food coloring, using a different color for each ball. You will need one red, one yellow, one green, and one orange (mix red and yellow food coloring to make orange).

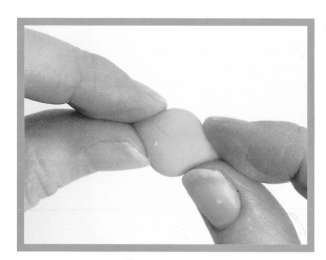

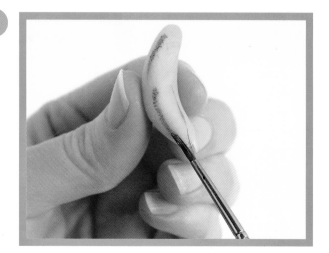

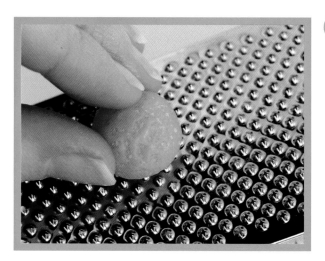

4 Knead the color into the marzipan until it is well blended and the color is even. You are now ready to make the fruit.

5 For lemons: roll small pieces of yellow marzipan into lemon shapes and roll over a fine grater to give the surface texture.

6 For bananas: take small pieces of the yellow marzipan and roll into a curved banana shape. Use a clean, fine paintbrush to paint lines onto the bananas with brown food coloring.

7 For oranges: make small balls with orange marzipan and roll over a fine grater to give the surface texture. With a toothpick, make a small indentation in the top and put a dot of brown in to indicate a stalk.

8

9

10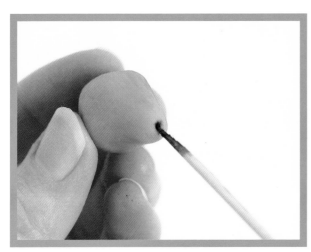

8 For strawberries: mold small pieces of the red marzipan into strawberry shapes. Roll over a fine grater to give the surface texture. Take tiny pieces of green marzipan, flatten into small leaf shapes, and press into the base of each strawberry to make leaves.

9 For apples: use red or green marzipan to make small apple shapes. Slightly indent the top of each apple and, with a toothpick, put a dot of brown in to indicate a stalk.

10 For pears: take small pieces of green marzipan and mold into pear shapes. With a toothpick dipped in brown, make a small indentation in the top of each pear to indicate a stalk.

St. Clement's Creams

Makes about 30

YOU WILL NEED

1 orange, skin scrubbed clean
1 lemon, skin scrubbed clean
1 cup powdered sugar, sifted
4 teaspoons glycerin
about 30 orange and lemon
    fruit jelly candy pieces

1 Finely grate the rind of the orange and set aside in a small dish. Now finely grate the rind of the lemon and set aside in a separate dish.

2 Squeeze the juice from the lemon and set aside. Then squeeze the juice from the orange and set aside in a separate dish.

3 Divide the powdered sugar equally between two mixing bowls.

4 Add the lemon rind,
2 teaspoons of lemon juice, and
2 teaspoons of glycerin to one
bowl and mix well. Now add
the orange rind, 2 teaspoons of
orange juice, and 2 teaspoons
of glycerin to the other bowl
and mix well.

5 Sprinkle a surface with
powdered sugar and knead the
orange mixture and the lemon
mixture separately.

6 Dust a tray and your palms
with powdered sugar. Using
one flavor at a time, pull off
small pieces of the mixture and
roll into a ball between your
palms. Place each ball on the
prepared tray.

7 When you have
used up all the
mixture, press an
orange jelly
segment (for the
orange mixture)
or a lemon jelly
segment (for
the lemon
mixture) into
each ball and
leave for
several hours
to firm up.

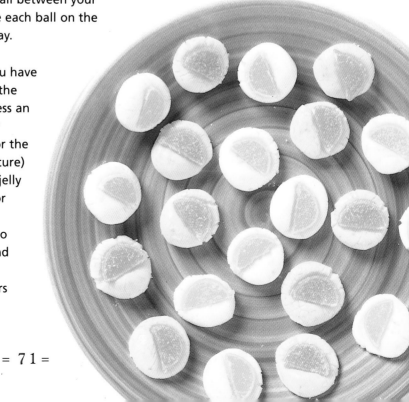

Honey Fruit Balls

Makes about 20

YOU WILL NEED

¼ pound ready-to-eat dried dates
¼ pound ready-to-eat dried figs
¼ pound ready-to-eat dried peaches
1 tablespoon honey
¼ cup sesame seeds

1 Place the fruit on a chopping board and **chop finely**.

2 Put the chopped fruit into a mixing bowl and stir well to combine.

3 Add the honey and mix again.

4 Take heaping teaspoons of the mixture and roll into balls between your palms.

5 Put the sesame seeds in a small dish and roll the fruit balls in the sesame seeds to coat.

CHEF'S TIP

☞ Instead of sesame seeds, try coating the Honey Fruit Balls in finely chopped nuts.

Coconut Snowballs

Makes about 30

YOU WILL NEED

1 cup powdered sugar, plus
  extra for decoration
⅝ cup sweetened condensed
  milk
½ cup dessicated coconut

1 Sift the powdered sugar into
a large bowl.

2 Add the condensed
milk and stir to combine.

3 Add the dessicated
coconut, stir, and
then turn out onto
a board. Knead
the mixture until
well combined.

= 74 =

4

5

6

4 Pull off small pieces of the mixture and form into small balls. You can leave them with a coarse, roughed-up surface or roll them between your palms to give a smooth surface. Or try a mixture of the two.

5 Sprinkle with a little powdered sugar for a snowy effect and leave to firm up on a tray.

6 When firm, arrange the snowballs in small paper cases.

Crunchy Clusters

Makes about 35

YOU WILL NEED

1 cup cornflakes
¼ cup chopped mixed nuts
¼ cup raisins
2 tablespoons sesame seeds
¼ stick unsalted butter, cut into pieces
1 tablespoon corn syrup
¼ pound baker's chocolate, broken into small pieces
¼ cup filberts, coarsely ground or chopped

1 Put the cornflakes into a plastic bag and crush lightly with a rolling pin. Put the crushed cereal, nuts, raisins, and sesame seeds into a bowl and mix well.

2 **Melt the butter and syrup in a small pan over a gentle heat, stirring occasionally. Remove the pan from the heat** and place on a heatproof surface. Add the chocolate and stir until melted.

3 Add the cereal
mixture to the melted chocolate mixture and stir
until everything is coated in chocolate.

4 Put teaspoons of the mixture into small paper
cases.

5 Sprinkle with ground or chopped nuts and
leave to set.

Rainbow Suckers

Makes 8

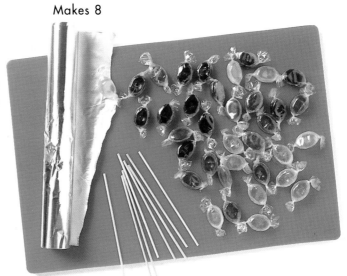

YOU WILL NEED
3 ounces red hard candy
3 ounces yellow hard candy
3 ounces purple hard candy
You will also need: extra-
 strength foil and eight
 wooden skewers

1 **Preheat the oven to 350°F.** Using a bowl or plate approximately 5 inches in diameter, mark eight circles on foil, and **cut out**.

2 Press each foil circle into a patty pan approximately 4 inches in diameter and run your finger around the inside bottom rim to indent a circle in the foil. Make sure the overlap is sticking up all around.

3 Remove each foil case from the patty pan, place on baking sheets, and insert a wooden skewer at the base of each foil case.

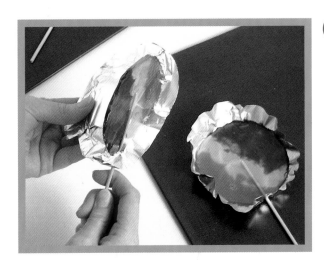

4 Place each group of hard candy in a separate plastic bag and **crush with a rolling pin**.

5 Put a thick layer of crushed candy in each foil case, arranging the different colors into three stripes.

6 **Bake in the preheated oven for about 3 minutes** until the crushed candy has melted. Wearing oven gloves, **remove the baking sheets**, place on a heatproof surface, and leave to cool.

7 When completely cold, peel the foil away from the suckers.

Jolly Jello

Makes about 1½ pounds

YOU WILL NEED

1 x 15-ounce can of fruit in
juice, such as apricots,
cherries, strawberries
1 pound sugar, plus extra for
tossing
⅝ cup very hot water
1 ounce powdered gelatin

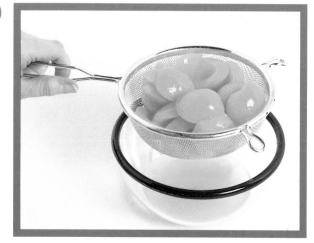

1 Drain the juice from the fruit, put the juice in a
saucepan, and the fruit into a bowl.

2 Mash the fruit and then purée it. Add the
purée to the juices in the saucepan. You should
have about 1 cup of liquid in total. **Warm the fruit
liquid, but do not let it boil. Remove the pan from
the heat,** add the sugar, and stir it to dissolve.

3 Put the hot water in a cup or small bowl and
sprinkle gelatin over it.

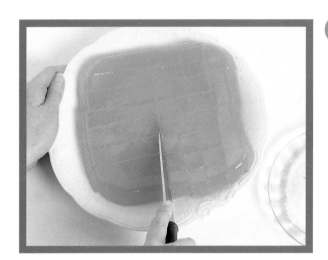

4 Stir with a teaspoon until the gelatin has dissolved completely and is no longer gritty. If the gelatin does not dissolve properly, **stand the cup in a saucepan of just-boiled water (taken off the heat and placed on a heatproof surface) and stir** until it does dissolve.

5 Add the gelatin mixture to the fruit liquid and stir thoroughly.

6 Pour into a shallow 7-inch square pan and leave to cool. Then place in the fridge for several hours to chill and set.

7 When set, dip the base of the pan into a bowl of hot water for a few seconds to loosen. Place a wet plate over the pan and flip the two together so that the jello falls out onto the plate. **Using a sharp, wet knife, cut into cubes.** Toss in sugar. These cubes should be eaten fairly quickly.

Crispy Crackles

Makes about 36

YOU WILL NEED

¼ pound marshmallows
¼ cup candied cherries
½ cup puffed rice cereal
⅛ cup dessicated coconut
4 tablespoons unsalted butter

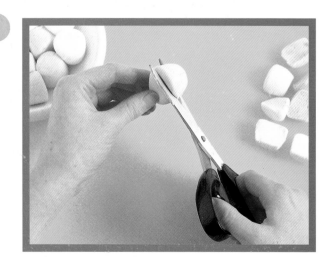

1 Line a square 7-inch pan with foil. If using large marshmallows, **cut them in half using scissors dipped in powdered sugar**. Set aside.

2 **Chop the candied cherries** into small pieces.

3 Put the cherries in a bowl with the puffed rice cereal and coconut and stir to combine.

4 Put the marshmallows and butter in a large saucepan. **Melt the marshmallows and butter over a gentle heat, stirring continuously.**

5 When the marshmallows and butter have just melted, **remove the pan from the heat** and place on a heatproof surface. Stir the cereal mixture into the saucepan and mix well to combine.

6 Put the mixture into the prepared pan and press down with the back of a spoon. Leave to cool and set for two hours.

7 When set, turn out onto a chopping board, remove the foil and **cut the crispy crackles into squares**. Place in small paper cases to serve.

CHAPTER FOUR

Perfect Presents

Mini Florentines

Makes about 18 cookies

YOU WILL NEED

3 tablespoons corn syrup

4 tablespoons unsalted butter

¼ cup light brown soft sugar

¼ cup all-purpose flour, sifted

½ cup flaked almonds

¼ cup chopped filberts

½ cup candied cherries, chopped

¼ pound semisweet chocolate chips

1 **Preheat the oven to 350°F.** Line two baking sheets with parchment paper.

2 Put the syrup, butter, and sugar into a saucepan and **stir over a gentle heat to melt. Remove the pan from the heat** and place on a heatproof surface.

3 Add the flour, nuts, and cherries and stir well to combine.

4 Put teaspoons of the mixture onto the parchment-lined baking sheets, leaving plenty of room for spreading.

5 **Bake for 7 minutes in the preheated oven until lightly browned.** Wearing oven gloves, **remove the baking sheet from the oven**, place on a heatproof surface, and leave the cookies for 10 minutes. With a palette knife, remove the cooked Florentines from the baking sheet and leave to cool on a wire rack.

CHEF'S TIP

☞ You can make bigger Florentines. Put larger spoonfuls of the mixture onto the baking sheets and let them cook for a little longer – about nine minutes.

6 Meanwhile, **bring a saucepan of water to the boil. Remove from the heat** and place on a heatproof surface. Place a bowl over the top of the pan, making sure the water does not touch the bowl. Put the chocolate chips into the bowl, leave to warm up, and then stir until it has melted. Remove the bowl from the pan.

7 With a small palette knife, spread chocolate onto the bottom of each Florentine and place on a wire rack, chocolate side up, to dry.

8 When the chocolate has started to set, mark wavy lines in the chocolate using the prongs of a fork.

Marvelous Macaroons

Makes about 16 cookies

YOU WILL NEED

enough rice paper to line two
   baking sheets
2 egg whites
¾ cup powdered sugar, sifted
½ cup ground almonds
few drops almond extract
¼ cup flaked almonds

1 **Preheat the oven to 325°F.** Line two baking
sheets with rice paper.

2 Put the egg whites into a very clean bowl and
beat with an electric beater until stiff.

3 Gently but thoroughly fold in the powdered
sugar, ground almonds, and almond extract with a
large metal spoon.

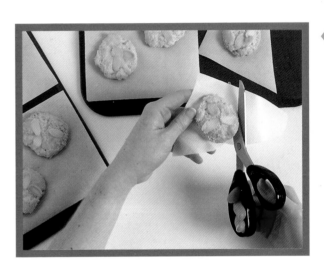

4 Put teaspoons of the mixture onto the baking sheets lined with rice paper, leaving plenty of room for spreading.

5 Place about three pieces of flaked almond in the center of each macaroon.

6 **Bake in the preheated oven for 20-25 minutes** until golden. brown. Wearing oven gloves, **remove the baking sheets from the oven**, place on a heatproof surface, and leave until cool enough to handle.

7 **Cut the excess rice paper from the macaroons** and leave on a cooling rack until cold.

Vanilla Walnut Fridge Fudge

Makes about 1½ pounds

YOU WILL NEED

1 pound powdered sugar, plus
   extra for sprinkling
1 stick unsalted butter
½ cup chopped walnuts
1 teaspoon vanilla extract
3 tablespoons heavy cream

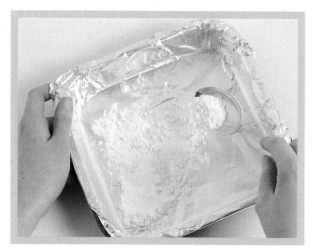

1 Line a square 7-inch pan with foil and sprinkle
withpowdered sugar.

2 Sift 1 pound of powdered sugar into a bowl.

3 <u>**Melt the butter in a saucepan over a gentle
heat.**</u>

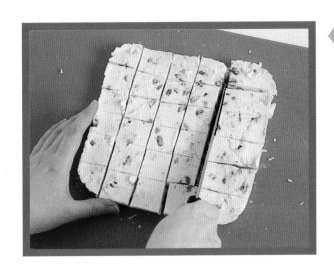

4 Add the melted butter to the powdered sugar and stir with a wooden spoon to combine.

5 Add the remaining ingredients and mix thoroughly.

6 Spoon the mixture into the prepared pan and smooth down with a palette knife, pushing the mixture into the corners of the pan. Leave to chill in the refrigerator for several hours.

7 When set, turn the fudge out of the pan and peel off the foil. **Mark the fudge into squares and cut with a knife.**

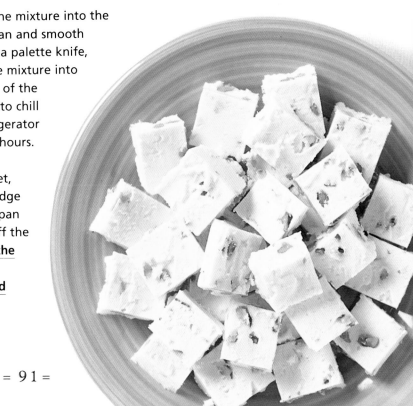

Minty Melting Moments

Makes about 40 mints

YOU WILL NEED

¾ pound powdered sugar, plus
  extra for kneading
6 teaspoons glycerin
2 teaspoons water
½ teaspoon peppermint
  extract
pink and green food
  coloring

1 Sift the powdered sugar into a large bowl.

2 Add the glycerin and 2 teaspoons of water and mix well.

3 Add the peppermint extract and work into the mixture. The result should be a stiff paste. Add a little extra water if necessary.

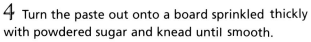

6 Dust a surface and a rolling pin with powdered sugar and roll out each piece of paste separately to a thickness of about ¼ inch. Using a small round or decorative cutter, cut the paste into shapes, mark with the prongs of a fork, and leave for several hours to harden.

4 Turn the paste out onto a board sprinkled thickly with powdered sugar and knead until smooth.

5 Divide the paste into three and color a third pink and a third green. Leave one third white.

Yummy Choc Truffles

Makes about 24 truffles

YOU WILL NEED

½ pound baker's chocolate, broken into small pieces

½ stick butter, cut into pieces

⅝ cup heavy cream

1 cup powdered sugar, sifted

¼ cup multicolored sugar strands

¼ cup chocolate strands

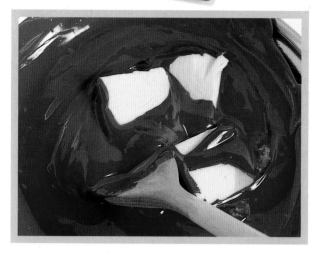

1 **Bring a saucepan of water to a boil**. Remove from the heat and place on a heatproof surface. Place a bowl over the top of the pan, making sure the water does not touch the bowl. Put the chocolate into the bowl and stir until it has melted.

2 Add the butter to the bowl and stir to melt. Remove the bowl from the pan.

3 Add the cream and powdered sugar to the bowl and whip with a wire beater until the mixture is smooth. Pour it into a shallow pan and chill in the refrigerator for several hours.

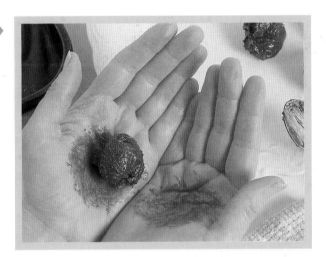

4 When firm, scrape up teaspoons of the mixture and roll it into balls with your hands. This can get a bit messy, so try to keep your hands cold (so the chocolate does not melt too much) and keep a clean, damp cloth nearby to wipe your hands if necessary.

5 Spread the sugar strands on two plates, one for each type. Roll half the chocolate balls in colored sugar strands and the other half in chocolate strands. Place each truffle in a paper case and then into a pretty box.

CHEF'S TIP

☞ Truffles should be kept in the fridge and eaten within four days.

Measuring Charts

LIQUID MEASURES

Standard	Metric
¼ teaspoons	1.25 ml spoon
½ teaspoon	2.5 ml spoon
1 teaspoon	5 ml spoon
1 tablespoon	15 ml spoon
1 oz.	25 ml
2 oz.	50 ml
2½ oz.	65 ml
3 oz.	85 ml
3½ oz.	100 ml
4 oz.	120 ml
4½ oz.	135 ml
5 oz.	150 ml
6 oz.	175 ml
7 oz.	200 ml
8 oz. (1 cup)	250 ml
9 oz.	275 ml
10 oz.	300 ml
12 oz.	350 ml
14 oz.	400 ml
15 oz.	450 ml
16 oz. (2 cups or 1 pint)	475 ml
18 oz.	500 ml
20 oz. (2½ cups)	600 ml
1½ pints	750 ml
2 pints	1 liter
2½ pints	1.2 liters
3 pints	1.5 liters
3¼ pints	1.6 liters
3½ pints	1.7 liters
4¼ pints	2 liters
4¾ pints	2.25 liters
5¼ pints	2.5 liters
5¾ pints	2.75 liters

SOLID MEASURES

Standard	Metric	Standard	Metric
¼ oz.	10 g	14 oz.	400 g
½ oz.	15 g	15 oz.	425 g
¾ oz.	20 g	1 lb (16 oz.)	450 g
1 oz.	25 g	1¼ lb	550 g
1½ oz.	40 g	1½ lb	675 g
2 oz.	50 g	2 lb	900 g
2½ oz.	65 g	2½–2¾ lb	1.25 kg
3 oz.	75 g	3–3½ lb	1.5 kg
3½ oz.	90 g	4–4½ lb	1.75 g
4 oz.	100 g	4½–4¾ lb	2 kg
4½ oz.	120 g	5–5¼ lb	2.25 kg
5 oz.	150 g	5½–5¾ lb	2.5 kg
5½ oz.	165 g	6 lb	2.75 kg
6 oz.	175 g	7 lb	3 kg
6½ oz.	185 g	8 lb	3.5 kg
7 oz.	200 g	9 lb	4 kg
8 oz.	225 g	10 lb	4.5 kg
9 oz.	250 g	11 lb	5 kg
10 oz.	300 g	12 lb	5.5 kg
11 oz.	325 g	13 lb	6 kg
12 oz.	350 g	14 lb	6.5 kg
13 oz.	375 g	15 lb	6.75 kg

OVEN TEMPERATURES

Fahrenheit	Centigrade
225°	110°
250°	130°
275°	140°
300°	150°
325°	160°
350°	180°
375°	190°
400°	200°
425°	220°
450°	230°